NCC
The National Computing Centre

The National Computing Centre develops techniques and provides aids for the more effective use of computers. NCC is a non-profit-distributing organisation backed by government and industry. The Centre

- co-operates with, and co-ordinates the work of, members and other organisations concerned with computers and their use

- provides information, advice and training

- supplies software packages

- publishes books

- promotes standards and codes of practice

Any interested company, organisation or individual can benefit from the work of the Centre by subscribing as a member. Throughout the country, facilities are provided for members to participate in working parties, study groups and discussions, and to influence NCC policy. A regular journal – 'NCC Interface' – keeps members informed of new developments and NCC activities. Special facilities are offered for courses, training material, publications and software packages.

For further details get in touch with the Centre at Oxford Road, Manchester M1 7ED (telephone 061-228 6333)

or at one of the following regional offices

Belfast   1st Floor
             117 Lisburn Road
             BT9 7BP
Telephone: 0232 665997

Birmingham   2nd Floor
             Prudential Buildings
             Colmore Row
             B3 2PL
Telephone: 021-236 6283

Bristol   6th Floor
             Royal Exchange Building
             41 Corn Street
             BS1 1HG
Telephone: 0272 277077

Glasgow   2nd Floor,
             Anderston House
             Argyle Street
             G2 8LR
Telephone: 041-204 1101

London   11 New Fetter Lane
             EC4A 1PU
Telephone: 01-353 4875

Designing Systems for People

Leela Damodaran
Alison Simpson
Paul Wilson

PUBLISHED BY NCC PUBLICATIONS

British Library Cataloguing in Publication Data

Wilson, Paul
　　Designing Systems for people.
　　1.　Electronic data processing
　　2.　Computer engineering
　　3.　Man-machine systems
　　I.　Title
　　001.6'4　　　　　　　　QA76
　　ISBN 0-85012-242-2

First published in 1980 by:

NCC Publications, The National Computing Centre Limited, Oxford Road, Manchester M1 7ED, England

Printed by H Charlesworth & Co Ltd, Huddersfield

Typeset in 10pt Press Roman by Focal Design Studios Limited, New Mills, Stockport, Cheshire

ISBN 0-85012-242-2

Acknowledgements

This handbook is a collaborative venture between the National Computing Centre (NCC) and the research group on Human Sciences and Advanced Technology (HUSAT) of Loughborough University of Technology. In the course of writing this book the authors have used material supplied by members of both these organisations past and present. The principal contributors are as follows:

Eric Clark
Anne Clarke
David Coan
David Davies
Ken Eason
David Hebditch
Leon Heller
Brian Pearce
Susan Pomfrett
Brian Shackel
Tom Stewart

The authors' thanks are due to the people and organisations listed below. They have contributed to this study by participation in formal interviews, by reading and commenting on the draft book and by informal discussion of opinion and experience, all of which have helped to shape the final form of this publication.

Apex Word Processing Working Party
Barclays Bank International
IBM
ICL
Metal Box Limited
National Water Council
Phoenix Assurance Company Limited
Post Office Research Centre

Tony Wight
David Pascall
Dr John Axford
Colin Hampson Evans/Bill Morton
Mike Bone/Richard Elliot
Don Weald
Graham Briscoe
Roy Yates

Sun Alliance Insurance Ray Walker
Tesco Limited Peter Sturgess

The Centre acknowledges with thanks the support provided by the Computers, Systems and Electronics Requirements Board (CSERB) for the project from which this publication derives.

Preface

'Designing Systems for People' — surely we do that already, so why *another* book?

Increasingly computers are being used *directly* by many different types of people. On-line interactive networks, remote entry and response facilities, and of course desk-top systems are all putting computer power straight into offices and other workplaces rather than leaving it in some distant data processing centre. The 'users' are no longer only the programmers, dp staff, and specialists who have developed computing expertise; many computer systems must in future be designed for 'non-expert' users.

More emphasis must now be placed on the man-machine interface and on all aspects of communicating between systems and people. Otherwise, inadequate facilities are provided: computers are used inefficiently and the users may well become bored, tired and error-prone. The ergonomic aspects of terminal selection and use must become part of the system designer's portfolio of expertise.

These ergonomic aspects are presented in this book in as broad a context as possible. However, there remain many other ergonomic and wider human science issues. Computer system designers will not become ergonomists or human scientists by using this book, but they do need to have available the relevant experience and advice from such human scientists, and this has been our aim. This book represents the results of collaboration between computer specialists (staff of the National Computing Centre) and human scientists (from the HUSAT Research Group in the Department of Human Sciences at Loughborough University). The NCC is well-known for its educational and methods-study work on systems design.

This book does not intend to teach systems analysis and design nor the basic concepts of human science: references are given to some good sources for such information. The aim is to combine principles from applied ergonomics with precepts from practical design experience, for systems designers to use and adapt to their needs.

Since the book aims to provide specific advice, its recommendations must be interpreted according to the reader's requirements. Allowance must be made for on-going technological change, and the effects of organisational, social, cultural and other 'real-life' influences must also be considered. Some of the content may influence, and will certainly be influenced by, company policy considerations.

Do not take this book as idealistic or directive. It is intended to be a practical guide. We believe it will be found useful in the general world of computer-based system design, and it is meant to be used imaginatively by thoughtful analysts and designers.

D R A COAN
National Computing Centre

B SHACKEL
Department of Human Sciences
Loughborough University of Technology

Introduction

The provision of effective systems to meet organisational objectives is the primary objective of computerisation. In the early days of commercial data processing, the emphasis in system design was on solving technical problems. This related to the limitations of hardware and software and to the technical problems of computerising a business function. A computer-centred approach developed out of the unavoidable preoccupation with the limitations of the technology.

In recent years the reliability, flexibility and power of computers have increased dramatically. Also there has been a growing realisation of a gap between the predicted benefits of computerisation and the actual achievements. Users have reacted to computer data processing constraints by partial use, misuse and non-use of system facilities. In addition there is a growing awareness amongst users that computer technology can lead to a restructuring of work patterns. The computer-centred approach has been challenged by these developments.

During the last 10 years it has become apparent that to fully exploit computing capability requires a more user-centred approach to systems design: the technology now permits a high degree of adaptation to the needs of the non-expert user. It is clear that new skills and knowledge are needed by systems designers and analysts.

The needs and characteristics of humans in regard to systems design — the human aspects of systems design — represent a very broad field of human knowledge. It is also a complex area since humans vary, with differing experience, skills, hopes, fears and expectations. Therefore, the guidance given in this book must be adapted to suit particular users and environments.

Computing operations have been traditionally based mainly on centralised computers to which input documents were sent and from which output was received some days later. This is rapidly changing: on-line links to computers are now commonplace, thereby allowing users to communicate with the machine and to receive an almost instantaneous response via computer terminal equipment. The introduction of terminal equipment into the user environment has been partly responsible for the upsurge in recognition of human aspects problems. Hence a large part of this book deals with problems specific to the selection, installation and use of terminal equipment.

Many types of terminal are in use and more are being developed. This book makes detailed recommendations for only the most widely used terminal — the visual display unit and associated keyboard. However, the *general* issues and principles raised by VDUs and keyboards are likely to be of equal importance for other types of terminal.

The cost benefits of considering human aspects are difficult to assess. Introducing a new technology into a work situation has an impact on organisational structure, methods of work, and the design of jobs. Frequently these changes occur as a by-product of increasing automation and are often ignored. Evidence of the long-term adverse effects upon motivation, job satisfaction, communication, work load and career prospects reveals considerable hidden costs in the medium-to-long term. These costs are incurred when users fail to respond positively to the new technology. Slow learning of new procedures, high error rates, lowered output and varying degrees of passive or active resistance represent significant financial losses. Such costs, which are not itemised in any cost benefit analyses, relate to a complex array of human factors issues. These issues are of far-reaching significance, both in humanitarian and financial terms.

There is a further, purely pragmatic, reason to take account of human aspects in systems design: an increasing number of trade unions are beginning to negotiate new technology agreements with employers. Such agreements effectively preclude employers introducing new technology such as computer terminal equipment without first negotiating with the union on the human aspects of such technology. It is hoped that this book will help to ease the path of such negotiations by providing a comprehensive and easy-to-use guide which is of equal use to both employer and trade union.

This book addresses itself to activities which constitute successful systems development. Some of the human aspects covered in this volume are not traditional systems design activities, eg design of jobs and workstations. However, a varying number of different types of specialist do get involved, both formally and informally, in the process of systems development. Such people include not only systems and programming personnel, but also people such as business analysts, O & M staff, personnel staff, trade union officials, training staff, heating and ventilation engineers and lighting engineers. This book attempts to integrate those aspects of these various functions which affect the user. In particular, the book aims to help the systems designer.

Contents

1 Planning the Development

SETTING UP LINES OF COMMUNICATION

Introduction

The development of a new computer system is likely to result in considerable changes to the working environment and will be of interest to anybody involved in that particular environment. If details of the proposed development are not available, rumours may spread, giving rise to misconceptions and unjustified fears. Therefore effective and timely dissemination of information about the development is essential.

The following groups of people should be kept informed about a particular systems development:

— users;

— management;

— unions;

— user associates.

Users

Users should be informed by their own management as early as possible about any systems development that will affect them, before colleagues elsewhere hear. Any announcement may well awaken fears about job security, changes in job content, and loss of discretion or influence. Therefore, where possible, the announcement should include positive statements about such issues, as well as clearly stating the purpose of the development.

It may also be useful to explain why commercial organisations need to change in order to prosper, and to indicate the many alternative ways of organising work which have been successfully attempted in other industries. Presentations, discussion sessions and films provide useful ways of getting such points across. Before the development gets under way all users should be given a rough idea of the timescale and the various stages of the project. In addition, users should be quite clear about the role they are to play, if any, in the development.

All users should be kept aware of general progress and anticipated completion dates of the project. This can be achieved by departmental meetings, printed news bulletins or by a user representative on a steering committee or belonging to the design team.

Management

User management will obviously be well-informed about the project, but other management in the organisation should also know of its existence in case they have dealings with the area concerned, and so that they are aware of the current deployment of the organisation's resources. Management meetings and brief printed bulletins are appropriate media for informing such managers of the purpose and progress of the development and anticipated completion dates.

Unions

In a trade union environment, union members are likely to ask their officials to ensure that the development will not be detrimental to themselves or their jobs. It is probably best for unions to be informed as early as possible about development projects: there is little point in implementing plans if all must come to a grinding halt for negotiations. Union officials are likely to be particularly interested in matters such as job losses, job changes or the operation of new equipment. It is sensible to include at least one union official on the steering committee. If detailed negotiation has to be undertaken, existing industrial relations machinery can be used or a steering committee subcommittee set up.

User Associates

People who have regular dealings with the users may well be affected by the development. Such people need to take the new system into account in their own plans, and to anticipate changes in the user department's attitudes. General information about the purpose of the development, current progress and anticipated completion dates will be sufficient for this group of users. Printed bulletins can be used for this purpose.

PLANNING THE DESIGN STRATEGY

Introduction

Few systems designed for a known user population (ie the employees of an organisation) are designed without user involvement. Traditionally a team of computer systems experts take responsibility for system design and implementation, and invite user involvement as they deem appropriate. Reasons for seeking user involvement include:

— *local knowledge*: to create an effective system it may be necessary to develop a detailed understanding of the task environment within which the system must operate. Only the potential user population may have this knowledge in detail;

— *resistance to change*: many systems have been rejected or poorly used because users felt the system was being imposed upon them. It is widely believed that involving users in systems design reduces the chances of resistance to change when the system is implemented;

— *user demands*: computer systems can lead to widespread changes: jobs can be lost; skills can become irrelevant; sources of power can be eroded, etc. Many employees, and especially their union representatives, are now demanding the right to examine the systems design plans;

— *increasing industrial democracy*: throughout the western world there is a move towards industrial democracy. Increasingly, it is considered right and appropriate for employees to be involved in computer systems design.

The different routes to user involvement lead to different forms of involvement. Underlying the different approaches is the question of responsibility for systems design (see below).

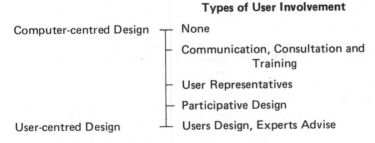

Figure 1.1 Types of User Involvement

No User Involvement

There are some disadvantages in user involvement. User involvement may be seen as unnecessary because the design team contains all the expertise required, or because involvement is judged to be costly and impractical within the time and resource constraints of the project.

The first argument relates only to the need to tap the local knowledge of users and may be valid where systems designers have extensive experience as users. However, it is increasingly felt that users have the right to be involved in decision processes which may affect their future.

The second argument about the cost of user involvement is more germane. It is certainly true that it can be costly and because it is a difficult process to manage effectively there is no guarantee of success. Groholt (1978), for example, reports that user involvement has rarely been successful because neither designers nor users have access to suitable procedures. However, far from suggesting the abandonment of user involvement, Groholt advocates more far-reaching forms of involvement. Some form of user involvement should be attempted although it must be recognised that success may be limited. In most cases the direct costs of user involvement will be more than balanced by the hidden costs of non-involvement.

One-Way Involvement: Communication, Consultation and Training

These techniques are under the complete control of the design team. Various means (newsletters, meetings, noticeboard announcements, etc) can be used to keep users abreast of systems developments but they all tend to limit the user response. These techniques are useful for increasing system acceptability but they cannot be expected to lead to active commitment and support.

Consultation is taken here to mean the process by which designers might tap the expert knowledge of users by conducting some form of task or job analysis. This process enables the user to make a valuable contribution. However, the user is usually constrained by the question he is asked and may be unable to voice worries.

Finally, training is an essential part of system implementation but it is one-way (from trainer to trainee) and occurs too late for the user to influence design decisions. These techniques each have a valuable contribution to make and from the designer's viewpoint are relatively safe and straightforward to conduct. Where users have serious doubts about the wisdom of systems design solutions, however, these techniques are unlikely to allow those doubts to be heard.

User Representatives

When the aim is to ensure that user opinion is heard effectively within systems design, a very popular strategy is to involve user representatives in the design process. This can be accomplished in the following ways.

Representatives in the Design Team

Employing users to supplement the technical experts in the design team has a number of advantages. Designers have user expertise close at hand whenever they need it and if the user is trained in systems techniques, the design team effectively expands its manpower. As many organisations have discovered, user representatives can also perform very useful liaison roles between users and the system after implementation.

There are also disadvantages. A technical designer might favour this arrangement because it provides safe and controllable user expertise within the design team. However, the user will be influenced more by other designers than by his user origins and will become progressively less representative of users. Indeed, if the user has volunteered for full-time membership of the design team, he probably intends a career in systems design and is not a typical user.

Moreover, with user representatives close at hand, the design team may not contact and consult other users. If they get the wholehearted support of representatives they may mistakenly believe that they have the support of all users.

One approach is for the representative to be a part-time member of the design team and to retain contact with the users he represents, but this too has its problems. Representatives find themselves caught between equally convincing but incompatible arguments by users and designers. It needs maturity and understanding by everyone to help representatives cope with such dilemmas.

Finally, many user representatives who are not trained in systems techniques find it difficult to understand design drawings and flowcharts and impossible to understand the system ramifications. Similarly they will have difficulty in expressing user needs and knowledge in a way that is useful to technical designers. This is a general problem of all methods of user involvement; the right medium of involvement exists but not the methods for effective medium use.

Representatives on Consultative or Steering Committees

User involvement in the design team is primarily useful in the detailed specification of the system. There remains the question of assessing the wider implications of the proposals for users' jobs and the organisation. These implications are more often examined within steering committees which manage system projects and frequently include user representatives. At this level users are more able to see the strategic issues associated with system proposals and they may be in a position to challenge unacceptable consequences.

Experience with this arrangement has raised a number of issues. Many users were members of consultative committees which had no executive control over the design team; their views tended to be ignored when they were contrary to design-team objectives. Users are now tending to seek committees with direct executive powers.

A second issue is the membership of the committee. There is a need for all sections of user population to be represented but frequently senior members of departments are expected to represent junior staff, and affected departments which are not central to the design are not represented. The question of conflicting loyalties is again an issue. Representatives will find themselves explaining the systems design case to their colleagues and being accused of joining the 'other' side.

A further problem occurs if the committee member is officially given no time to consult the people he represents. The user may represent no-one's views but his own; whilst he may become enthusiastic about the project, uninvolved users may remain apathetic or hostile.

Finally there is often negotiation in deriving an acceptable system solution. Questions of redundancy, retraining, loss of job skills, demarcation, etc, may have to be settled before the system can be implemented. Most organisations have formal negotiating machinery for dealing with questions of pay and conditions: the relationship between this machinery and a project steering committee can be a problem. The existing machinery is usually incapable of dealing with the broader social and organisational implications of design proposals. *Computerisation Guidelines* (NCC, 1979) recommends that companies and unions review their consultation and negotiation methods to assess their relevance to computer technology.

Participative Design

One-way involvement and user-representative techniques cannot adequately utilise the expertise, nor gain the commitment, of all of the user population. To counteract this, various forms of participative design are now being employed which seek to bring *all* users into the design process. Not all users should be involved in programming, technical systems design, etc, but they should be involved in selecting the form of work organisation they will operate, ie the jobs they will do, how tasks will be allocated between them, and the task relationships that will exist between employees and the computer system. It is only after decisions are taken at this level that the technical system specification can be formulated. Thereafter, as the technical system is developed, users are consulted on detailed issues associated with the achievement of the agreed specification.

When this approach is successful, users greet the system with enthusiasm and commitment. Inevitably, however, there are problems. It means that technical systems designers play a minor role in the early stages of the project because it is the users who specify the system requirements. It is difficult to manage a design process when there are many users, and the most successful applications tend to have been with small groups of users.

The central problem is again technique: users need considerable help to conceive of different forms of work organisation and to consider which is most appropriate to them. Participative design therefore requires people who are able to provide information and concepts which are useful without precluding users from making their own decisions. There can also be problems with user attitudes to this process. When an organisation has hitherto not involved employees in decision-making, they will inevitably be suspicious of the motives for soliciting involvement.

A scheme can have a successful beginning and then can go badly awry when computer systems designers seek to implement the agreed specification. The problems of meeting user requirements can lead to designers radically changing the specification. In one case, design problems led to a radical change proposed by employees and accepted by management and designers being subsequently rejected in favour of a system which supported the status quo. In work at Loughborough University, it has been found that a continuous process of trials, consultations and pilot schemes is necessary to prevent this kind of drift occurring.

Users Design: Experts Advise

Normally, senior management give the power and responsibility for design to the technical experts who thereafter may involve users as they consider appropriate. Because this has frequently led to users finding their requirements ignored or rejected, a number of more radical proposals have been made. These proposals have one feature in common: power and responsibility for systems design firmly reside with the users, and the computer technology experts are cast in the role of advisers. The critics of this approach point out that it frequently leads to a re-assertion of technical control whenever problems occur and user needs are ignored.

Legislation

Legislation is contributing to this shift in power. In countries where industrial democracy is already enshrined in law there are rapid movements in this direction. In Norway, for example, an employee has the legal right to be involved in the design of any computer system that is going to affect his work. Also, there are a variety of laws in Sweden which affect systems design.

Union Action

There is growing recognition amongst unions that computer technology has widespread implications for their members, and they are making major efforts to protect the interests of their members in systems design. Within some unions there now appears to be a major division of opinion about how this may be accomplished. One union view is that this can be done by working within the current framework, eg joining steering committees, etc, and participating in design in order to promote user interests. The other view assumes that management and worker requirements are in conflict, and that by joining a structure dominated by the management requirements the needs of the workers will not be met. Hence some unions are demanding that the users should control the design process or alternatively, the union should assemble technical experts to complement the experts employed by managers. Systems design would then become a process of negotiation, with each side supported by an expert team. The degree to which this strategy gains support will no doubt be inversely related to the degree to which users find they are able to participate successfully in systems design.

Deciding the Design Strategy

The appropriate degree of user involvement needs to be decided. In practice this will be influenced by prevailing management style, the industrial relations climate, attitudes of employees, and attitudes of systems personnel. In some areas, such as health and safety, involvement of users or their representatives could be a requirement negotiated as part of a new technology agreement.

The Need for Human Factors Expertise

A requirement for human factors expertise will exist irrespective of the degree of user involvement in the design process. There is much relevant knowledge of hardware characteristics and the physical environment. Design criteria are presented in this book to aid equipment selection and environment design.

Particular areas may require additional specialised expertise. For example, design of the software interface requires knowledge of dialogue design (image quality, methods of text presentation, formatting).

General guidance is given here for the systems designer to improve the software interface from the user's point of view. However, experimental design, for example, requires special expertise.

In 'social technology' areas, such as planned change and the design of jobs, the book can only indicate some of the relevant problems and choices. In these areas the diagnostic skills are as important as the design capability, and it could be unreasonable to expect the system designer to possess all the necessary skills.

PLANNING FOR DISRUPTION

Introduction

The normal running of a workplace can be disrupted in different ways during the process of a computer development. It is vital that this is realised and planned for well in advance. The first step towards coping with the problem is simply ensuring that everyone involved is aware of what disruptions are likely to occur. The most common causes of disruption are:

— management control activities;

— user contributions to systems design;

— physical changes to the workplace;

— installation of equipment and furniture;

— training the users to operate the new system;

— system testing, parallel running and system initiation.

Management Control Activities

Management needs to undertake the following additional activities:

— formal project control and progress activities;

— assessing design proposals;

— managing the disruption caused to users and operations by the development.

It may be necessary to reduce a manager's normal work load to permit adequate participation in the development process.

User Contributions to Systems Design

The amount of time the user will have to spend on the design work will vary depending on the design strategy that has been chosen. This type of disruption often takes the form of system designers visiting and interviewing users throughout the project. Initial visits are likely to be relatively formal events to acquaint the designer with the user's job. Later visits may be quick and informal to clear up specific points. Whatever the design strategy, the user's normal pattern of work will be disrupted to a certain extent. User management and users should be given prior warning.

Physical Changes to the Workplace

It is sometimes necessary to change the structure of the room to accommodate computer terminal equipment or new working arrangements. For example, the office may be split into two in order to create a room to house a printer; or two adjoining offices may be made into one larger room. In such instances the normal daily routine of the office occupants will be interrupted, if not halted, for some considerable time.

It may also be necessary to change certain visual features of the environment. For example, light sources may need to be altered, and curtains or blinds fitted or changed. Redecoration may continue over a long period of time (for example the repainting of all walls, ceilings, etc). Moving the people concerned to another location during this period may be the best solution. The disruptions may be reduced by moving people to different parts of the room, or by installing temporary screens to reduce dust and noise. The people affected are more likely to by sympathetic if they know every effort is being made to deal with the problem. Involving the people concerned is likely to increase their tolerance.

Installation of Equipment and Furniture

The lead times of furniture and equipment can be extremely long and should be checked before ordering to ensure that items arrive when they are needed. Furniture or equipment should not be installed until actually required; otherwise it may be seen as a hindrance. If some items are to arrive earlier than the rest, this should be planned for: they should be stored until required.

The date on which new equipment or furniture is to be installed should be advertised at least a week in advance. People can then plan their schedules and work accordingly. If the date has to be changed, all those who will be affected should be informed. If new furniture is delivered, the old furniture should be removed quickly. If new desks have been ordered the users will have to move all their belongings and this can cause complications. Management should ensure that each person has enough time, bearing in mind normal duties.

Training the Users to Operate the New System

The user is likely to have to undergo a thorough training programme, which may involve learning how to use the equipment, learning about the logic of the system, and about its implications for the day-to-day work routine. Such a training programme can be demanding on a user's time.

System Testing, Parallel Running and System Initiation

During this period users have a number of problems. First they may have to run the old system *and* the new one until the old system is abandoned. There is also the extra work involved in identifying and correcting system bugs, and it will take a few weeks for users to become familiar with the new system.

Management should ensure that users understand their responsibilities during this period and that they can cope with their work load.

2   Systems Analysis — User Aspects

INTRODUCTION

If human aspects are to be taken into account in system design, particular analysis work is essential. This must take place *before* normal systems analysis procedures. (It is assumed the reader is familiar with the latter and will have had some experience in their application.) This preliminary analysis should involve:

— work situation appraisal;

— user identification;

— task demands;

— work role analysis;

— user readiness.

There is no single, standard technique for investigating the social or human system. Widely-used methods include observation, interviews, critical incident techniques, diary-keeping and variance analyses.

WORK SITUATION APPRAISAL

This preliminary process represents an 'intelligence phase', the major objective being to gain a general understanding of a work situation before specific details are sought. Where the systems analyst has experience of the particular work environment, this may involve renewing acquaintances and up-dating knowledge of the work situation.

Where the systems analyst is less familiar with the situation, he needs to obtain information. Given the broad objective of 'gaining understanding', detailed procedures cannot be specified. It is necessary to appreciate the ethos, customs and practices, as well as the work process and its pressures and priorities.

Success in this phase of analysis relies heavily upon establishing a rapport with the future users such that they volunteer their views, opinions, hopes and fears. Informal discussions, often focussed upon topics entirely divorced from the system, and contacts made away from the pressures of the work place can help the understanding of the work situation as seen by the various users.

Work Situation Appraisal: Aims

— to give a clearer idea of the specific information to be collected and of the relevant collection methods;

— to give the new systems analyst some background knowledge of the company, its products, and primary functions together with a knowledge of the business functions and the identity of individuals involved with them;

— to identify the necessary routes to follow through the organisation structure in the detailed analysis phase. People who have been employed by an organisation for some time will have some understanding of the different departmental functions. However, a deeper understanding can only be gained from experiencing a situation first-hand or from in-depth discussion with those who live with the problems and issues of day-to-day operations.

Conducting the Work Situation Appraisal

This process is partly unstructured, in sharp contrast to the subsequent detailed analysis procedures. Also, it needs to be 'non-threatening' to the users. It is impossible to predict the specific utility of some items of information gained. Part of the knowledge acquired may not have any direct application, and may only become useful much later in the design process.

The systems analyst's increased sensitivity to the users' perspective will help him to avoid designing systems which could be highly unacceptable to the user population.

Finally, it may be difficult for the systems analyst to build an effective relationship with users on account of differences in age group, sex, social or ethnic background, etc. In such situations, it may be appropriate to nominate an individual other than the systems analyst to gain the relevant broad-based knowledge.

In one successful application, a member of a management services team was selected as an acceptable liaison officer on the basis of being of the same sex, age and general background as members of the primary user group. Her role was defined in such a way that she could contribute important information concerning users to the system design team. In addition she gained feedback from users on systems proposals as they were generated. This approach proved very effective and highly acceptable to all concerned. The important point is that the systems analyst needs information but may not be the best person to obtain it.

One way of approaching the appraisal is to see its target as being to complete a 'stake holder' analysis. Each system will affect a variety of people – managers, clerks, customers, the public, even computer staff – and each may be considered to have a 'stake' in the kind of system developed. A 'stake holder analysis' is a two-stage operation:

i) Identify everyone who has a stake in the system (ie all those who will be affected directly or indirectly);

ii) Identify the nature of the interest of each group and make a preliminary assessment of their likely reaction. If the intelligence phase is conducted sufficiently early, several different versions of the system may still be under discussion. In this case the stake holder analysis can be used to assess the way different groups are likely to respond. This can help planning which system to implement and how each user group should be involved in design. Figure 2.1 is a schematic outline of the end-product of a stake holder analysis.

Stake Holder	System Plan 1		System Plan 2		System Plan 3	
	Advantages	Disadvantages	Advantages	Disadvantages	Advantages	Disadvantages
Dept A Management Section 1 Section 2						
Dept B Management Section 3 Customers Systems Staff etc						

Figure 2.1 'Stake Holder' Data Sheet

USER IDENTIFICATION

Users of Any Existing System

It is reasonable to assume that the users of the old computer or manual system will operate the new. The names and job titles of these users should be identified and recorded, and can provide the basis of a 'map' of the future which represents in diagrammatic form the location of the various users.

This 'map' identifies primary, secondary and tertiary users, and will establish from the outset that the system will not be used exclusively by a 'closed' group of central users. Although a few systems are purpose-built for the exclusive use of a small number of users, most commercial systems have many users who use only a few facilities.

Computer systems often affect more people than is obvious at first. Computer printout frequently crosses departmental and even organisational boundaries. 'End-users' may even be members of the public. These various users can become excessively demanding if their needs are not recognised and reflected in the development phase. It is therefore helpful to recognise that 'hidden users' may have comparable design and support requirements to those of the primary users.

A typical example of the hidden user is the company auditor who may audit the system on a regular basis. Another example is the senior manager who is given verbal reports from his subordinates about information produced by the system. In this case the senior manager is a secondary user who is being supported by a human interface who operates the system on his behalf. As a third example, members of the general public are often indirect tertiary users by virtue of receiving computer-produced bills or by purchasing tickets via an organisational representative.

TASK DEMANDS

The jobs of individuals are designed to achieve particular goals. Currently, systems analysis tends to concentrate upon those tasks which will become computer-based and therefore seeks to identify recurrent and quantifiable features of tasks.

From a human factors viewpoint, one requirement is to describe the functions of the total man-computer system. The computer system will perform certain functions but the human system will have to cope with everything else. If the computer system is intended to support and augment the activities performed by people, the functioning of the human system should be examined carefully.

Activities which will be left to the human 'social' system will be concerned with:

- abnormal, unexpected and unusual varieties of activities;
- data that is ambiguous, uncertain or unpredictable;
- data that cannot be easily presented in computer-compatible form;
- the resolution of conflicts between goals, and goal priorities (eg trading off short- and long-term aims);
- goals that cannot be expressed in quantitative terms (eg satisfaction aesthetics, aspects of quality, etc);
- a need for novel solutions (eg in design, problem-solving and decision-making);
- a need to interact with other people during the activity (eg airline seat reservations).

The above functions will be performed by people and it is therefore important that the computer system should support them. For example, if a system is designed such that it requires a fixed sequence of activities on the part of the user, difficulties arise if the user needs to perform them in a different order. In such a situation, the system must be designed to permit variation in the way the user works. Similarly, knowledge of the human/social system will identify software design issues and evolutionary issues which relate to how fast the total man-computer system may need to change in order to succeed.

WORK ROLE ANALYSIS

Users' perceptions of computerisation are frequently coloured by the way users believe their work roles will be changed by the proposed development. Work role refers to the set of behaviours and activities which characterise a given job.

Individuals will often place higher order interpretations on the global function of their jobs than might be expected from an analysis of the component tasks in those jobs. Thus a job function described by an analyst as a set of routine clerical procedures to facilitate paper-work may be seen by the job holder as 'helping the public' or 'assisting the sales group'. These interpretations are important human factors considerations because they indicate the value the job holders place on their work.

The information on the work role should be a major consideration in job design. The impact of a computer system can then be channelled to enhance and support important aspects of the work role rather than replace them with additional routine activity.

Work role analysis conducted with sensitivity and skill can reveal important motivational factors in the user's job and therefore provides base-line data for considering job design. It can provide advance warning of the positive and negative attitudes that may be expected towards the planned system. In particular, such analyses should indicate whether cost/benefits are likely to be wrong for the individuals concerned. For example, computerisation can affect career prospects, job content, etc. Work role analysis can be conducted to discover the rewards and costs of existing jobs by asking questions such as:

— what parts of the job are rewarding and/or satisfying?

— what aspects of the work are frustrating and/or dissatisfying?

— what career and promotion prospects exist currently?

The information gained through work role analysis will be helpful in determining where computer aid is needed most, which aspects of work should remain unchanged as far as possible and in achieving effective job design.

USER READINESS

It is essential to identify the extent to which users are ready to cope with and accept a system. Without such an appraisal, even the 'best' system can be rejected for reasons such as:

— the users do not have the skills to use it;

— opportunities do not exist for acquiring the requisite skills;

— users are suspicious of the new technology.

Once the user population has been mapped, it is important to consider such issues as: why should the user be concerned with the system? Is the user going to be dependent on the system? If he is going to be dependent on it, then it is appropriate to consider the type of required training and user support. If the user is not dependent on the system, it will have to be very accommodating since he can otherwise ignore or reject it.

Individuals will compare the perceived benefits of the system against the costs of their personal effort and time. The value of the system for his job and career progression will be assessed by the user.

The degree of choice facing the user will vary widely. It may be possible to coerce people into using an unsatisfactory system when their only alternative is to leave the company. With senior managers and members of the general public, coercion is rarely practical.

In designing training and user support schemes, the existing knowledge and skills of users need to be compared with the requirements of the computer system. The products of these forms of analyses have direct reference to:

— types of dialogue design;

— types of support provided.

To acquire this information requires some form of stimulation to which people can react. In particular, reactions to system proposals can reveal a great deal of valuable information. It becomes evident very quickly whether the demands being made on potential users are seen by them to be reasonable.

3 Equipment Selection

EQUIPMENT SURVEY

Introduction

This section introduces equipment that facilitates communication between humans and computers. This survey should be used to build a state-of-the-art picture of the subject. New developments can be noted as they occur.

The ways in which computers accept input and provide output are very unlike the means whereby human beings exchange information. The basic mismatch between people and computers arises primarily because of engineering problems: it is impossible to make reliable machinery which can match the flexibility of the human.

Human Communication Channels

The human being has several possible channels of communication by which he can receive or send information (Figure 3.1). In fact only a few of these channels are commonly used to achieve communication between humans and computers; for special applications, and at a price, the more unusual channels may be used. Figure 3.2 indicates the 1980 state-of-the-art for computer terminal (ie input/output) equipment.

Some of the more widely used items of equipment mentioned in Figure 3.2 are discussed below.

Visual Display Units

Three types of Visual Display Units (VDUs) were available at the beginning of 1980:

— alphanumeric VDUs;

— presentation graphics VDUs;

— vector graphics VDUs.

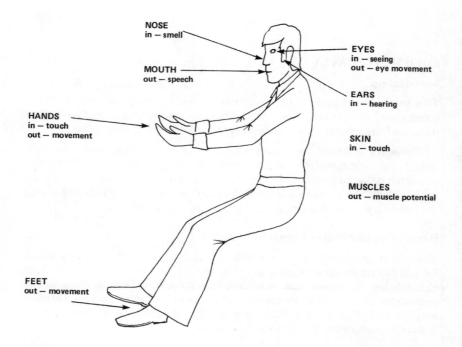

NOSE
in — smell

MOUTH
out — speech

HANDS
in — touch
out — movement

FEET
out — movement

EYES
in — seeing
out — eye movement

EARS
in — hearing

SKIN
in — touch

MUSCLES
out — muscle potential

Figure 3.1

Type of equipment	1980 state-of-the-art
Output from the computer	
Visual display units	Widely used. Several basic types exist with different manufactured versions.
Printers	As for visual display units.
Voice	Research prototypes available and a few manufactured versions.
Smell	Little work at present. Possible uses in fault detection.
Touch	Research on aids for the blind.
Input to the computer	
Keyboards	Widely used. Several basic types exist with different manufactured versions.
Character recognition.	A number of different manufactured versions exist. Increasing use being made of this technology.
Voice	Research prototypes exist. Some manufactured versions are available.
Finger/hand/ arm movement	Several basic types exist (eg light pens, digitizing panels, touch panels, joysticks) with different manufactured versions.
Leg movement	Little use made of this channel although the technology exists.
Eye movement	Research (mainly for the military).
Muscle Potential . . .	Research on direct electro-physiological control of artificial limbs and also on direct brain electric signal control of external units.

Figure 3.2 State-of-the-art Computer Terminal Equipment

Many different commercial versions of *alphanumeric VDUs* came onto the market in the late 1970s, giving a wide range of choice at competitive prices. There were signs that some manufacturers were beginning to produce ergonomically sound designs, and hopefully this trend will continue.

Presentation graphics VDUs are essentially alphanumeric VDUs which can display simple representations of data such as histograms or pie charts. When linked to a plotter which can assist in the production of overhead slides, such a system can be a powerful business aid. Colour, obviously a big advantage in presentation graphics, is fully available on many systems. This type of VDU, which emerged in the late 1970s, is likely to become more popular as costs diminish and experience grows.

The demand for full *vector graphics* facilities is limited to special applications, most of which fall into the category of design work. Applications at the Computer-aided Design Centre (Cambridge) have shown the value of good graphics systems for appropriate tasks. But developments in this field are still limited by costs, and the successful application of a vector graphics system often depends on the skill and enthusiasm of the user. However, UK research projects have aimed to improve the ergonomics of computer-aided design, a trend that is likely to continue. Research results should increasingly appear in the products offered in the marketplace.

Printers

A large spectrum of products was available at the end of the 1970s. At the high-output end of the market, very fast versatile laser printers could produce up to 20,000 lines per minute in a variety of fonts and with a graphics capability. However, such machines are rare: many smaller machines are finding their way into the workplace. Probably the major ergonomic problem with such machines has been noise, and now many manufacturers are beginning to combat this problem – either by developing quieter technologies or by incorporating silencer hoods into their products. This trend should continue as users and systems designers become more aware of the problem and learn to discriminate more when printers are purchased. By 1980 the market saw at least one satisfactory printer combined with VDU.

Voice Output

Research into voice output from computers dates from the early 1960s, and by the end of the 1970s a few commercial systems with limited vocabulary were available. Such systems appear to work reasonably well for applications such as enquiries to distant databases. Significant developments may be expected in the 1980s.

Keyboards

Keyboards are the most widely used technique for inputting information to computers. The design of computer input keyboards has been greatly influenced by traditional typewriter design. Consequently the QWERTY layout for typewriters has become the de facto standard for the computer input keyboards — despite the acknowledged inefficiency of the QWERTY layout and the many ingenious alternatives that have been proposed. The financial and skill investment in the QWERTY layout has prevented its replacement by more efficient designs. In addition, there are many special-purpose keyboards, eg the 10-key Numeric Keyboard and the Chord Keyboard. The 10-key *numeric keyboard* comes in two basic layouts (one for telephones and the other for calculators) as shown in Figure 3.3.

CALCULATORS			PUSH-BUTTON TELEPHONE		
7	8	9	1	2	3
4	5	6	4	5	6
1	2	3	7	8	9
	0			0	

Figure 3.3 10-Key Numeric Keyboard Layouts

Both layouts are of the '3 x 3 rows plus 1' type. Evidence shows that this arrangement is preferred to a layout of 2 rows of 5 keys, and that layouts other than 3 x 3 + 1 or 2 x 5 should be avoided if possible.

The 3 x 3 + 1 layout with numbers 1, 2, 3 in the top row (telephone layout) has been shown to be preferable to that with 7, 8, 9 in the top row (calculator layout) for accuracy of human performance. Moreover, it has been found that the accuracy advantage of the telephone layout is maintained whether the user has been trained for that layout only, or has been trained on both and alternated between the two. There is little difference between the two layouts in terms of speed of use. For the 2 x 5 layout the numbers should be 1 to 5 in the top row and 6 to 9 plus zero in the bottom row.

The basic principle of *chord keyboards* is that a single character is obtained by pressing two or more keys, thereby reducing the number of keys required and the size of the keyboard. It is claimed that chord keyboards offer a very ¡ast, truly portable alternative to the typewriter. However, the increased learning time they usually require is a disadvantage. Although work on chord keying was extensive in the early 1960s, relatively little more was done until the late 1970s when the stimulus of cheap hand-held calculators, and the obvious possibilities of hand-held data entry and 'writing', yielded a number of prototypes. At least one model was on the market at the end of 1979 and it is clear that the state of the art will advance rapidly in the early 1980s.

Character Recognition

Considerable work has been done on methods by which marks or characters can be read by computers, and various commercial systems are available. There are two basic types of character recognition equipment:

- alphanumeric recognition;
- mark recognition.

Equipment is available which will read *printed* alphanumerics with 100% accuracy. However, by early 1980 a satisfactory system for the computer reading of human hand-written alphanumerics had not been produced. A number of systems were available but they all constrained human handwriting in some way, eg requiring each character to be written in a single box and in capitals only. However, such systems have shown themselves to be very useful and it is likely that they will become increasingly common in the 1980s.

Alphanumeric recognition is usually achieved by optical scanning (optical character recognition — OCR), although some late-1970s products used press-ure-sensing devices to detect the writing of a character. Another alphanumeric recognition technique employs magnetic ink (magnetic ink character recognition — MICR), a technique mainly confined to banks.

It is easier for computers to recognise marks than hand-written alphanu-merics. Consequently, by the late 1970s a number of companies were offering Optical Mark Sense (OMS) equipment. Such systems usually require short lines to be entered on a preprinted form. Quite a high degree of accuracy can be achieved by this method, with 100% accuracy achievable by using printed marks (as, for example, in a sales ticket system). In the international bar code system every product is given a unique bar code which can be automatically read in warehouses, distribution systems, shops and at checkout counters. The use of a similar system for issuing and checking books in public libraries has already become widespread.

Voice Recognition

Voice recognition research has been going on since the mid-1960s, but by the end of the 1970s there were few effective commercial systems on the market. However, a cheap and versatile voice-input device would certainly challenge the keyboard as the most widely used piece of input equipment, a circum-stance that should encourage development through the 1980s.

Light Pen

The light pen is primarily a graphic input device. It is used in conjunction with a keyboard, or a menu on the display screen, to identify the displayed item to be processed. In office systems, light pens can be used either to indicate choices (probably in a menu or tree structure), or to directly manipulate data on the screen (eg moving pieces of text or displayed data). In some systems only a light pen is needed but usually the required alphanumeric input demands a keyboard. The main problem with light pens is their use on a vertical or near vertical screen. This can be extremely fatiguing for the wrist and arm. In normal offices, it is not practical to tilt the screen as light reflections obscure the display.

Digitizing Tablets

Digitizing tablets are generally used for inputting engineering drawings into computer systems, but can also be used for entering other types of data. The position of a cursor on the tablet is detected by electronic, electromechanical or ultrasonic means. Small tablets may use a pen-like device for the user to point; in some versions a modified pen may be used on a suitable overlay. Overlays have been developed for use in, for example, stock-control systems where the pictorial representation on the tablet helps the user to identify products.

Touch Panels

Touch panels are similar to digitizing tablets but identify the user's finger position rather than a stylus. If a touch panel is placed over a VDU screen, identifying items on the screen becomes a natural pointing process. However, the additional surface may cause reflections and can be obscured by finger-marks.

Joysticks

Joysticks and trackerballs (or rolling balls) may be moved or rotated to move cursors about a screen in the same direction. Both are widely used in Computer-Aided Design (CAD) and military applications and could presumably be used in office systems. By early-1980 many people were becoming accustomed to using joysticks in home television games.

Mouse

The mouse is a small device with two wheels at right angles to each other which code mouse movement in the X and Y directions. It can therefore be used to move a cursor around a display screen. In an experiment at the Xerox Alto Research Centre, the mouse was found to be faster and more accurate than a joystick or normal cursor control keys in selecting text on a VDU screen.

CRITERIA FOR EQUIPMENT SELECTION

Introduction

When selecting any computer terminal equipment a list of criteria must be drawn up against which each model must be evaluated. Three different types of criteria need to be satisfied:

- functional criteria;
- general criteria;
- human-aspects criteria;

Functional Criteria

Functional criteria specify what the equipment must be able to do (eg reading of handprinted data, printing of graphs in 4 colours, storage of 20,000 keyed characters for up to 14 hours, and automatic transmission overnight). These criteria can be established by assessing the activities allocated to both user and computer, and by inferring what characteristics the equipment will have to possess to enable those activities to be carried out; and also by being aware of the type of technology that is currently available.

General Criteria

General criteria include such things as:

- delivery time;
- simplicity of installation;
- reliability;
- maintenance support;
- compatibility;
- supplier factors (including the stability and credibility of the supplier: is the product range to be developed? is the hardware of proven design?).

These are standard criteria to be considered when making any equipment purchase.

Human-Aspects Criteria

These specify such considerations as:

- workplace characteristics of the equipment so that they match the physical requirements of the human user;

— equipment controls and operating activities, so that the user will find them logical, consistent and easy to understand;

— equipment-display physical features and screen formats, etc, so that the user will find them logical, consistent and easy to understand;

— environmental factors (even though they are seldom under direct control or easily changed) because special environments place special constraints upon choice (eg in outdoor usage, flame-proof equipment in operating theatres, etc);

— documentation which aids installation, maintenance, etc;

— training and support from the manufacturers to assist in system installation, implementation and modification.

Human-aspects criteria can be built up from a detailed analysis of the functional criteria. To successfully define the criteria often requires the application of detailed ergonomics knowledge and experience. It may well pay to have a specialist do the job.

A list of some of the more important human-aspects criteria for VDU screens and keyboards is given in Appendix 2.

MECHANICS OF THE SELECTION PROCESS

Introduction

Selection of any item can easily be influenced by many irrelevant factors. For instance, a person may reject one manufacturer's product because the salesman is unacceptable: or may buy another's because of the manufacturer's advertising campaign, regardless of the product's actual qualities. This approach can result in costly mistakes and must be avoided in the selection process. The success or failure of a system can have an enormous impact on an organisation's business, and it is therefore essential to carry out a more objective analysis.

There are many methods of analysis, some based on a points-scoring system and others on a financial-scoring system. One of the simplest and most effective methods is the 'weighted ranking' analysis: here each item is marked (ie awarded points) against a list of predefined criteria which have been weighted according to their importance. This straightforward approach can be applied effectively to all items of computer terminal equipment.

The steps in analysis are:

A   draw up lists of criteria on which the evaluation is based;

B   identify *essential* criteria;

C   assign weightings to the *remaining* criteria;

D   draw up a list of factors for awarding scores to each criterion;

E   establish which companies manufacture the equipment you require;

F   evaluate each model;

G   compare results and make selection.

Steps A to D are best carried out by a committee of two or three rather than an individual, a committee being more likely to produce an unbiased view.

A Draw up Lists of Criteria

Three different types of criteria need to be satisfied when selecting computer terminal equipment:

– functional criteria;

– general criteria;

– human-aspects criteria.

Use section on Criteria for Equipment Selection to build a list of criteria which meets your own requirements. Ensure that each criterion is clearly defined to facilitate quantitative analysis. The committee should also be aware of criteria that are implicit in other criteria, so that each factor is not evaluated more than once.

B Identify Essential Criteria

Once the list of criteria has been drawn up, all *essential* criteria must be identified. If a piece of equipment does not satisfy each one of these essential criteria then it is totally unsuitable and need not be considered further. The list of essential criteria may be used to draw up a shortlist of suitable models which can then be evaluated further under the remaining *desirable*, but not essential, criteria.

C Assign Weightings to the Remaining Criteria

Once the essential criteria have been identified, each of the remaining criteria of a particular type (ie either *functional* or *general* or *human-aspects*) must be ranked against every other criteria of the same type. This is done by constructing a weighting matrix and then comparing, via the following scoring system, each criterion against every other:

Criterion A more important than B: Score 3

Criterion A equally important as B: Score 2

Criterion A less important than B: Score 1

The whole matrix, except the diagonal, should be completed, even though either side of the diagonal will be complementary. Complete the matrix by comparing the criteria along the top (as A) against those down the left side (as B) and totalling downwards. In this way the higher the weighting the greater its importance. Therefore the equipment selected will be the one with the highest final score.

Once the matrix has been completed and the scores totalled downwards, these scores can be used as weightings. It is more convenient, however, especially if there are many criteria, to convert these scores to weightings in the range of 1 to 10. To do this the following formula must be applied:

If criterion A has score X then its weighting will be calculated as:

$$\frac{10}{2(\text{no. of criteria} - 1) + 1} \; X - \text{no. of criteria} + 2$$

This must be rounded to the nearest integer to give the actual weighting, W.

This formula is derived from a generalised formula for moving the axis of a normal distribution.

D Award Scores to Each Criterion

The next step is to decide on what factors the hardware is to be judged for each criterion. These factors must be defined and assigned scores. The range 1 to 10 is usually satisfactory, where 10 is the best score and 1 the least favourable.

There are two types of scoring scheme that can be used, namely a *sliding scale* and a *fixed scale*.

Sliding scales are best where it is difficult to quantify the criteria and some kind of subjective judgment has to be made (eg where the designer assigns a number, according to assessment of a criterion from, say, 'excellent' to 'poor'). Where possible this type of scale should be avoided and measurable or identifiable factors used. *Fixed scales* are used where each criterion can be objectively defined (eg where the designer assigns a score according to the value of a criterion — say, price or size — in a range).

As all scoring schemes should be devised before selecting the models which are to be evaluated, it is possible that no model will score 10; indeed they may all be weak in a particular criterion and all score low marks.

E Establish Which Companies Manufacture the Equipment

The relevant manufacturing companies can be identified by perusing computer journals, attending computer equipment exhibitions or by obtaining up-
to-the-minute equipment surveys. (See Bibliography.)

The appropriate company can then be contacted and will be pleased to
demonstrate their equipment and discuss requirements. The essential criteria
identified earlier can now be used to eliminate some of the companies to
leave only a shortlist of equipment which can be evaluated further.

F Evaluate Each Model

Each manufacturer's model to be evaluated can now be analysed to see how
well it satisfies each criterion. Steps A to D must be complete before this
stage is embarked upon in order to eliminate bias.

For each criterion consider which factor the model best agrees with and
award it that score. When each criterion has been dealt with, multiply each
score by the weighting for that criterion and total these results for each type
of criterion.

G Compare Results and Make Selection

Once Step F has been carried out for each model under consideration and for
each type of criterion, the model with the highest score wins. If several
models have high scores which are close, or if different models win for each
type of criterion, then either the criteria can be refined, the weightings
changed, and the calculations done again, or additional criteria can be included or a judging panel can be assembled to make the final choice.

4 Systems Design — User Aspects

JOB DESIGN

Introduction

Job design defines the *tasks* which are the responsibility of an individual *job holder*. Computer systems change the tasks that individuals undertake. Where a system *adversely* affects jobs, even a good technical system can run into problems, especially:

— difficulties in implementing the system;

— user apathy and hostility leading to staff turnover, absenteeism, sickness, etc;

— usage of the system in unplanned ways or non-use of some facilities.

Jobs are often adversely affected by computer systems implementation, simply because job design is not considered a part of the system design process. In most cases a technical system is first designed and job effects are considered at the time of implementation. At this stage there is often very little choice about the form the jobs must take. In fact there are many ways of organising systems to yield jobs: some approaches produce jobs perceived as 'good' by employees, and some produce 'bad' jobs.

It is important that the system implications for jobs are considered early in the design process. This involves consideration of what constitutes a 'good' job, who decides, and how it is done.

Job Design Principles

The central principle is to provide a person with autonomy commensurate with responsibilities and abilities. Specific needs are recognised:

— the job should be reasonably demanding in mental terms and provide variety;

— the worker should be able to learn continuously on the job;

- the individual should have some minimal area of decision-making (ie some responsibility);
- the individual's efforts and accomplishments should be recognised in the workplace;
- it should be possible to relate work to life outside work (ie to feel that the job is significant and meaningful);
- it should be possible to feel that the job leads to some sort of desirable future (not necessarily promotion).

None of these needs is an absolute. The degree to which they are appropriate depends upon various factors, eg the history of the individuals, their expectations, and the expectations of employers.

Who Decides What Constitutes a 'Good' Job?

Decisions on job design should be taken in the light of local contingencies. To avoid false assumptions, decisions should follow the fullest consultation with the users and their management. It is sometimes possible to ask user groups to make their own decisions, with designers providing technical details. However, the extent of user involvement must depend upon local circumstances.

Establishing Alternative Job Designs

A fundamental problem is that no-one is accustomed to thinking of alternative ways of organising tasks into jobs. Thus there may be difficulties in seeing possibilities other than those embedded in the status quo. The first step therefore is to consider alternative strategies. This can be approached from three directions:

- *From the Literature.* The literature suggests various approaches to organising tasks into jobs. In *job rotation* there is regular staff movement from one task to another. This is useful if each task is boring; it adds variety and a wider perspective but the tasks need to relate to one another. In *job enlargement*, tasks can be added to the job to give more variety. This is sometimes known as *horizontal* job design. In *job enrichment*, tasks can be organised so that the user has responsibility for all the tasks necessary to complete a major sub-goal. This method adds substantially to autonomy and is sometimes known as *vertical* job design. Use can be made of *autonomous work groups*, where a group of related tasks constitute more work than one person could undertake.

- *From the Nature of the Work.* Work may be divided in different ways, each being used as a basis for constructing jobs. The possibilities depend upon the characteristics of the work but three are common. In *division by function*, the various clearly defined functions (eg collection of data,

data entry, interrogation of data base, printout checking and despatch, etc) are assigned to individuals. This is also known as *task specialisation*. It has the advantage of minimising training but it produces boring jobs and an inflexible structure in which it is difficult to re-allocate staff. Ignorance in groups of the activities of others also tends to produce problems of control and communication. Alternatively, in *division by product type*, it can often make an interesting job to handle all of the functions associated with a particular product. Here the job holder sees the process through from start to finish. There are no coordination problems and the job holder becomes an expert, able to give advice to customers, etc. In *division by customers, suppliers, etc*, the work originates from or goes to specific inside or outside bodies and it is possible to combine into one job all the tasks associated with one customer or supplier. This gives the job holder a strong sense of identity and value, and provides customers and suppliers with a clear contact point.

— *From Technical Determinism*. If job design is left to the end of systems design, the technology may preclude many options. It can do so in relatively trivial ways, eg the location of terminals may make it difficult to combine certain functions. A limited number of terminals may make it difficult to organise by product type, etc. These kinds of pressure tend to push job design towards a functional division (task specialisation), often the poorest form of job design.

Consider, for example, one common job design problem associated with computer systems. Most systems require data entry and it is often a clearly identifiable function. It is natural to think in terms of one or more persons having responsibility for data entry. However it can lead to:

— boring, repetitive work;

— high error rates because the data is meaningless;

— the operator being controlled by the machine, ie not a 'user' (psychologically a slave not a master);

— the operator being a full-time terminal operator with consequent danger of eye strain.

Against these consequences is the possibility of achieving a high keying rate, often a difficult aim to accomplish. Before adopting this strategy it is important to consider other ways of capturing the data which would permit the creation of more desirable jobs:

— collect the data at source, ie spread data collection over many more jobs;

— make it part of the transaction, ie capture it when it first appears;

— ensure each person deals with output as well as input, ie add an output function;

— ensure input and output have a perceivable relation, ie make the task meaningful. This can be done by a product- or customer-type of division;

— if central data entry is necessary, make the job more varied by adding other data-base tasks, eg help users with data-base queries.

The data entry example shows how technical system design can lead to job structures being determined without consideration of better alternatives.

Choosing Between Alternatives

This can involve a trade-off between a number of factors (some of which may be more difficult to quantify than others):

— cost, eg some solutions need more terminals than others;

— customer satisfaction and service;

— training requirements;

— operator satisfaction and efficiency;

— organisational flexibility;

— error rates and processing rates.

It is difficult to establish these factors and to determine the trade-offs. Debate is needed within the user population, and it is often useful to create a trial or pilot system as early as possible in design. Users may then be encouraged to experiment, to explore the implications of using the system in a variety of ways. An informed approach to good job design can then emerge.

ACTIVITY FEEDBACK

Introduction

Activity feedback means supplying information to a person about his activities (eg giving a salesman information about his sales). The case for activity feedback is clear: people learn and perform better if they are provided with feedback on their performance. Also people are more committed to their work, and to the organisation they work for, when they see how their work has contributed towards the organisation's performance.

Levels of Activity Feedback

The individual can be provided with several levels of feedback information. At the base level, information about tasks should be available. This information should contribute to the next level of activity feedback at, say, departmental level. These first two levels of information are important and should be provided wherever possible. They tell the individual how his work is progressing in the department.

Higher levels of activity information can also be provided. Departmental information should contribute towards, say, divisional information. This in turn could contribute to the top level of information describing the organisation's overall performance.

Such an information hierarchy has obvious parallels with a management information system. However, there are significant differences between the two:

- a management information system handles specialist information which allows managers to control the organisation, whereas an activity feedback information system deals with information that can be comprehended by most members of the organisation (eg number of machines sold, number of orders taken);

- a management information system uses abbreviations and jargon which are comprehensible to its limited range of users, whereas an activity feedback information system must use non-specialist terms;

- a management information system, unlike an activity feedback information system, is unlikely to concern itself with low-level information about individual activities.

Despite the differences, the relationship between the two types of system should always be considered when activity feedback information is being designed.

Designing Activity Feedback Information

Here, the following points should be remembered:

- the information should concern itself with activities that are meaningful to the intended recipients of the information;

- the information should relate to the objectives of the organisation;

- it should be possible to feed information at the base level into the next level;

- the information must be presented in a format and language suitable for the intended recipients;

- the information should be specifically relevant to the work;

- the amount of information supplied to an individual should be limited.

The aim is to provide a quick and straightforward assessment. Large amounts of information would complicate and confuse.

DIALOGUE DESIGN

Introduction

Dialogue design specifies the content and format of communication between users and computer systems. In the book, emphasis is placed on the design of dialogues using VDU screens and keyboard. However, many of the principles described can be adopted for dialogues conducted with other types of system.

The following information is needed before dialogue design can be undertaken:

- *Details of user activities*. This information is needed to identify the required dialogues and their objectives. The activities to be undertaken and the associated methods should have already been identified at the job design stage.

- *The roles of dialogue users*. This information is required for dialogue style selection. A clear picture of user roles should have been built up during job design.

- *Jargon, abbreviations, codes or cultural linguistic differences understandable to the user*. These should have been obtained during the analysis phase.

- *Details of the equipment*. A basic knowledge of the equipment is necessary to appreciate the dialogue constraints. For VDUs, for example, the designer must be aware of the screen size, the type of cursor employed, whether a reverse video facility is available, etc. The designer should read the equipment manuals, talk to people familiar with the equipment and use the equipment.

- *Details of existing dialogue design standards*. Any dialogue design standards that already exist in the organisation need to be applied in the design of any new dialogue. If such standards do not exist then the new dialogue should be compatible as far as possible with dialogues the users currently experience.

General Principles

Before embarking on the dialogue design process, a number of general principles should be appreciated:

- *Plan for considerable dialogue design effort*. It is easy to underestimate the effort required. Anticipate going through several draft designs and plan to try out each draft on the system to be used and on a representative group of end-users.

— *Apply dialogue design standards.* These standards can cover a wide range of dialogue facets, eg position of transaction codes on screen, signing on/off procedures, error messages, use of blinking or reverse video to make parts of the dialogue stand out. It is important to ensure that such facets are consistent across all dialogues encountered by a particular user. This can be a major problem when in practice the various dialogues the user encounters may have been designed by different analysts. Within a computer department each dialogue may be regarded as a completely independent entity within different systems, but to the user no such distinction exists. Therefore even if no dialogue standards are available, it is important to inspect other dialogues the user encounters and to ensure that the new dialogue is consistent with them. In addition, program function keys should have the same uses for each dialogue. This requirement might be relaxed if 'soft' keys, with the function displayed on the screen above each key, are available.

— *Design a flexible dialogue.* Design for a fluid population; even the same group of users will not have the same requirement of the system from one year to the next. Some users will learn quite quickly and wish to continue learning, others may have to start the learning process again due to absence. Design should be flexible from the beginning; then the system can change to meet new circumstances.

— *Design a logical dialogue.* The dialogue must be logical and easy-to-use. Every opportunity should be taken to obtain constructive contributions from the user throughout the design process.

Stages in Dialogue Design

The dialogue design process can be broken down into the following stages:

Identification of the data elements which are likely to occur in the course of the dialogue.

Selection of dialogue style.

Breakdown of the dialogue into discrete blocks produced alternately by the system and the user.

Dialogue design for each block, for users of intermediate capability.

Design of validation procedures and error messages.

Formatting the screen.

Testing and modification.

Designing flexibility into the dialogue to accommodate new users, very inexperienced users and/or other categories of users.

Identification of Data Elements

During the job design process, certain tasks will have been identified which require dialogue between user and system. Using this information, each self-contained dialogue should be identified, its objectives stated, and a name meaningful to the users given to it (eg 'sales order query' or 'sales order input'). Then all data elements likely to appear in a particular dialogue should be listed and an attempt made to define their characteristics. This is because one of the most difficult problems in dialogue design is knowing enough about each data element to cater for all the ways the user may wish to use it. Therefore the more information that can be amassed about each data element the better. In a database environment, data definition sheets may already exist for the data elements; if so, such sheets should be used in the dialogue design process. A comprehensive data definition sheet is shown in Appendix 3.

Selection of Dialogue Style

Choosing the dialogue style is probably the most crucial stage of the design process. The many different dialogue styles are all variations of eight main types:

- natural-language-based;
- programming-type dialogue;
- instruction and response;
- menu selection;
- displayed formats;
- form filling;
- panel modification;
- query by example.

Natural-language dialogue. It is possible to design a good natural-language-based dialogue. However, the problems of coding and data structures often lead to dialogues which are superficially 'natural-language-based' but which are in reality quite cumbersome and difficult to use. A well-designed natural-language-based dialogue can be suitable for many types of user because each can use the language (and language style) natural to him. An example of a natural-language-based dialogue is given in Figure 4.1. (Figures 4.1 to 4.11 are examples of dialogues as they appear on a VDU screen.)

In L R Harris's 'ROBOT' system the following inputs are acceptable:

1	LOS ANGELES AREA MANAGERS?
2	WHAT AREA MANAGERS LIVE IN LOS ANGELES?
3	WHICH ARE THE AREA MANAGERS WHO LIVE IN LOS ANGELES?

and would all be accepted as meaning:

PRINT THE NAME OF ANY EMPLOYEE WITH JOB = AREA MANAGER
AND CITY = LOS ANGELES

and the output would then be a list of names.

Figure 4.1 Example of Natural-Language-Based Dialogue

The programming style of dialogue is rarely relevant for any user type other than programmers and even then this style should not be used without very careful consideration and user consultation. An example of this type of dialogue is shown in Figure 4.2.

```
UPDATE PERSONAL FROM SOURCE 1;
STRUCTURE SEGOL FROM SEGB;
EQUATE;
EMPNO TO MANNO;
SALARY TO WAGE;
END EQUATE;
IF ACTION EQ 'D';
LIST PERSONAL RECORD;
REMOVE SEGOL;
ELSE;
DECREASE TAX RATE BY 5;
INCREASE BENEFITS BY 100;
IF ACTION EQ '1';
INSERT SEGOL;
IF ACTION EQ 'R';
REPLACE SEGOL;
```

Figure 4.2 Example of Programming-style Dialogue

Instruction and response dialogue. The example of this style of dialogue (Figure 4.3) shows how instruction and response can be used for infrequent users where the main task of the operator is to *input* data.

```
ENTER PERSONNEL CODE OR 'END'
1297684
EVANS LESLEY
ENTER SELECTION REQUIRED:        'PERSONAL'
                                 'EDUCATION'
                                 'EMPLOYMENT'
                          or     'SALARY'

SALARY
ENTER ITEM REQUIRED:             'BASIC'
                                 'GROSS-TO-DATE'
                                 'TAX-TO-DATE'
                                 'OTHER'

BASIC
BASIC IS £33000 PER ANNUM
DO YOU WISH TO CHANGE THIS ITEM? ENTER
'YES' OR 'NO'
NO
ENTER PERSONNEL CODE OR 'END'
END
```

NB UNDERLINED = INPUT

Figure 4.3 An Example of Instruction and Response Dialogue

Of course, this style of dialogue can also be used by experienced users, though in this case the prompts should be omitted or simplified as shown in Figure 4.4.

```
                                 PERSONNEL CODE
1297684
DAVIES GAIL                      PERSONAL
                                 EDUCATION
                                 EMPLOYMENT
                                 SALARY

SALARY                           BASIC
                                 GROSS-TO-DATE
                                 TAX-TO-DATE
                                 OTHER

BASIC
£37000 pa                        CHANGE?
NO                               PERSONNEL CODE
END
```

NB UNDERLINED = INPUT

Figure 4.4 Example of Instruction and Response Dialogue Without Prompts

To the experienced user, prompts could appear to be unnecessary. Instruction and response style dialogue is best suited to the work of clerks, the public and specialist users.

Menu selection is a technique that is often highly acceptable to inexperienced users. This technique is best used in situations where the data is very complex. An example of a menu selection dialogue is shown in Figure 4.5.

ELECTRO-CARS LTD							
SPECIFY ORDER DETAILS:		2, 4, 9, 12					
MODEL		MOTOR POWER		NO. OF DOORS		COLOUR	
1	SPARK	4	15 HP	8	2 DOORS	10	RED
2	SPARK DELUXE	5	20 HP	9	4 DOORS	11	GREEN
3	SPARK GT	6	25 HP			12	WHITE
		7	30 HP				
FIFTEEN HORSE-POWER WHITE SPARK DELUXE WITH FOUR DOORS							
PLEASE CONFIRM (Y/N): Y							

NB UNDERLINED = INPUT

Figure 4.5 Example of Menu Selection Dialogue

Experienced users often find this mode of interaction with a system rather unsatisfactory, and a modified menu method with users typing a few of the first letters of the chosen menu item without the menu itself being displayed can be a useful technique. The system can thus be menu-driven for beginners, but command-driven for experienced users.

Displayed formats are probably the simplest of dialogue styles and can be used efficiently in a wide variety of applications and on most terminals. In practice, the displayed format shown in Figure 4.6 could be omitted for experienced users who can remember the correct input order.

```
BOOK ORDER

ENTER AUTHOR/TITLE/PUBLISHER/ISBN/

NO. OF COPIES/CUSTOMER NAME/CUSTOMER ADDRESS/

POST OR COLLECT?

HATCH, GRAHAM/THEORY OF ECONOMIC GROWTH/MACMILLAN/
NOT KNOWN/18/RICHARD DAVIES/354 HIGH RD, HARROW
WEALD, HARROW, MIDDLESEX/POST
```
NB UNDERLINED = INPUT

Figure 4.6 Displayed Format Style of Dialogue

Form-filling dialogues involve displaying a format map on the screen which corresponds as closely as possible in layout to the related input document as shown in Figure 4.7. The 'map' is protected and cannot be altered by accident, by the user from the keyboard. The user keys data into 'variable' areas of the screen which are not protected. The data which has been entered can then be transmitted to the computer by pressing the 'send' key. Such techniques are very easy to use. With fixed-length fields, the cursor is automatically tabbed to the next input field when the last character is entered. With variable fields the user presses the TAB key when he has finished the input. This dialogue style works best when the user only inputs data and does not need to receive output extra to that displayed in the forms; complications arise when the computer has to display data in response to input.

```
NEW ACCOUNT DETAILS

ACCOUNT NUMBER (84978632) CUSTOMER ORDER REF (*      )

CUSTOMER NAME (DAVE EVANS            ) TYPE (BUTCHERS)

ROAD (18A ST ANN's ROAD*                                )

TOWN/CITY (HARROW*                                       )

COUNTY (MIDDLESEX*                                       )

DELIVERY ADDRESS (AS ABOVE*                              )

DELIVERY INSTRUCTIONS (DELIVER TO REAR OF PREMISES*  )
```
(= TAB STOP

) = AUTOTAB

* = DEPRESSION OF TAB KEY

NB   UNDERLINED = INPUT

Figure 4.7 Example of a Form-Filling Dialogue

Panel modification is only recommended for experienced operators handling complex data. Data is displayed on the screen in response to an input key such as 'salary details' (Figure 4.8).

NEXT FUNCTION TYPE			
SALARY DETAILS	WADE, PAUL MAXWELL		
FIGURES MARKED BY '*' MAY BE CHANGED			
BASIC SALARY	*47000	ANNUAL BONUS	*2000
GROSS PAY TO DATE	22600	TAX TO DATE	8700
PENSION FUND CONTRIBUTION	*02.50%		
TOTAL CONTRIBUTED TO DATE	579.00		
PENSION AT RETIREMENT	10500 PA		
NATIONAL INSURANCE RATE	2.45	*CONTRACTED OUT	
DATE OF LAST INCREMENT	31.01.85		

NB UNDERLINED = INPUT

Figure 4.8 A Panel Modification Dialogue

If required, fields may be modified and sent back to the system by users who have security clearance for that particular field.

Query by example is used when an enquiry or a search is made on a database. The data set involved is indicated by the user and the system responds by displaying the data elements in the record concerned (Figure 4.9).

EXAMPLE OF AN ENQUIRY QUERY

DATABASE:	PERSONNEL		
ELEMENTS:			
NUMBER 47863	NAME	DEPARTMENT	SALARY
NUMBER 47863	NAME CHATTERIS, CHRISTOPHER	DEPARTMENT PROMOTIONS	SALARY COMMISSION ONLY

EXAMPLE OF A SEARCH QUERY

DATABASE:	PERSONNEL		
ELEMENTS: NUMBER *	NAME *	DEPARTMENT SALES	SALARY 18000 — 22000
NUMBER 47402 47863	NAME SUTHERLAND, MARK WILSON, FRED	DEPARTMENT SALES SALES	SALARY 18450.00 21500.00
END			

NB UNDERLINED = INPUT

Figure 4.9 Query by Example Dialogues

Because this style is used for output of data, it is sometimes suitable for the public, managers and process controllers. Unfortunately, a simple paradigm for selecting a dialogue style has not yet been derived. Consequently the choice of a dialogue style must depend on the trial and error matching of task and user characteristics with the characteristics of the styles as described on the previous pages. To help in this process some of the more relevant characteristics of each style are summarised below.

Natural Language . . . . .Difficult to design successfully.
Suitable for most types of user.

Programming . . . . .To be avoided if possible.

Instruction and response. . . . .Suitable for the input of data by
infrequent users eg clerks, the
public and specialists.

Menu selection . . . . .Good for inexperienced users who are
handling complex data. Experienced
users will require a modified version.

Displayed format. . . . . .Suitable for infrequent users in most
applications.

Form filling . . . . .Suitable for input only.

Panel modification. . . . . . . .Suitable for experienced users handling
complex data.

Query by example . . . . .Best used for enquiries or searches on
data bases. Suitable for the public,
managers and process controllers.

It should be remembered that these styles are not necessarily mutually exclusive: a blend of styles can often be appropriate, especially if both experienced and inexperienced users have to be catered for.

Breakdown of the Dialogue into Discrete Blocks

A dialogue takes the form of alternate statements from the user and from the system. To start off the design process proper the objective of each statement, made first by computer and then by user, should be identified. These statements need not include error messages, requests for re-input, etc, since such issues will be dealt with later in the design process. Where possible, the identified objectives should then be ordered in the way they are likely to appear in the dialogue. The order of the objectives should form a progression which is logical and consistent to the user and should correspond to any order implied by the user's task, eg by the order of data on an input document.

Users should be asked for their constructive comments on the objectives that have been defined and the order they have been placed in. Any shortcomings that are thus identified will be far easier to rectify at this stage than after the design has been completed.

Dialogue Design for Each Block, for Users of Intermediate Capability

Each objective must now be taken in turn and the dialogue designed for it in the selected style. Where it is not possible to define an order for a group of objectives, a dialogue must be designed which will stand on its own, independently of dialogue coming before or after it. The dialogue designed at this stage should meet the needs of the intermediate user, as opposed to the experienced user who may need only a very abbreviated dialogue.

The guidelines under the following headings should be kept in mind in the course of this detailed design work:

— variations and limits;

— quick positive system responses;

— starting and ending dialogue;

— vocabulary;

— codes and abbreviations;

— brightness coding;

— reverse video coding;

— blink coding;

— colour coding.

Variations and limits. For each objective the variations and limits of the dialogue should be identified, eg for the objective 'input of date by user', variations could include: 12 January, 12/1, 80/1/12. For the objective 'output of percentage difference by computer', limits could be: minimum of 1 decimal point, maximum of 4 decimal points. Data definition sheets usually provide relevant variation and limit information.

Quick positive system responses. Each input by the user should result in some sort of quick positive response from the system, eg if the user has requested a search of a database which is taking some time, the system should say it is searching and, preferably, give an estimate of how long it will be before the information will be available. If a considerable delay in response is likely, it might be advisable to inform the user periodically that processing is still continuing.

Starting and ending dialogues. Blocks of dialogue must also be designed for starting and ending the overall dialogue. This may include signing on/off the terminal and undergoing certain security checks.

Vocabulary. The vocabulary used in the dialogue should be easily understood both by the user and by the computer. Unfortunately general prescriptions cannot be given for the vocabulary that can be understood by users because different users understand different words.

The most naive set of computer users is the general public and for such users 'computer jargon' should never be used as part of the dialogue. Some words will be meaningless and others will be misunderstood as shown by the examples given below.

bucket	either	–	in direct access storage a bucket is a unit of storage as distinct from the data contained in the unit
	or	–	something you put water in
buffer	either	–	generally used as a means of temporarily storing data when information is being transmitted from one unit to another
	or	–	what stops a train from crashing into a platform at a station
bus driver	either	–	a device for amplifying output signals
	or	–	someone who drives a bus
key	either	–	a digit or digits used to locate or identify a record, but not necessarily attached to the record
	or	–	something used to unlock a door
program	either	–	a set of instructions composed for solving a given problem by computer
	or	–	the American word for something watched on television

(Computer definitions reproduced from *A Dictionary of Computers,* P J A Chandos, Penguin, 1970.)

Sometimes such vocabulary problems can be compounded by the situation in which the system is to be used. In each situation the designer must find out how much 'computer jargon' the users readily understand.

When designing for the public, it is best for the dialogue to allow the computer to output, and accept input, in common English. However, this should not be done when designing for other types of user. The results of the work situation appraisal should be used to establish the technical jargon employed by the potential users every day at work (as shown by the example below).

In the glass-making industry,
cullet = broken glass
frit = sand

Wherever possible, this technical language should be used and *not* the equivalent in common English.

Codes and abbreviations. If alphanumeric codes or abbreviations are used in the manual system then these should be used in the computer system (however illogical they may appear to the system designer). The existence of such codes or abbreviations should have been identified during the work situation appraisal.

If new codes or abbreviations have to be produced, then these should be logical and unambiguous to the user. Therefore it is sensible to ask the potential users for suggestions and to test the proposed abbreviations or codes on the users to discover if they have any difficulty in identifying the derivation.

There are three basic methods used to produce codes:

— copying;

— association;

— transformational.

For the address:

'333 Leicester Road, Loughborough'
COPYING may produce "LOUGHBOROUGH"
ASSOCIATIVE may produce "241"
TRANSFORMATIONAL may produce "333 LRL" (where the rule is to use the number followed by initial letters of the remainder of the address).

Copying refers to the situation where a part of the full message is copied to become the code.

In *associative* coding a code is allocated to the message and thereafter associated with it. This code may appear to be totally unrelated to the message.

Transformational coding is where a standard rule is applied to the message in order to produce the code. The important thing is to be consistent, eg the same rule should be adhered to throughout a single task. When a complete list of abbreviations/codes has been produced, it should be examined for inconsistency, ambiguity and possible confusion. A list stating the full meaning of codes/abbreviations must be easily accessible to all operators either through the computer system or on look-up cards at each terminal.

There are further coding considerations:

— *Brightness coding*: where particular items of information can be made to stand out on the screen by displaying them at a different luminance level. To ensure the effectiveness of brightness coding no more than two levels of brightness should be used, corresponding to normal and bold appearance.

— *Reverse video coding*: another technique of making data stand out on the screen by displaying data which is normally shown as light characters on a dark background, as dark characters on a light background. This is a very effective method of making data stand out though it should be noted that where a large area of the screen is displayed in reverse video, flicker is more likely to be perceived.

— *Blink coding*: this attracts the attention of the user to particular items of information by blinking the character or field that represents the relevant data. Although blink coding is effective, it may be a source of annoyance to the user. Therefore it is best to limit blink coding to small fields and to provide the means of suppressing the blink action once the data has been located and is being attended to.

— *Colour coding*: an effective means of visual coding and enhancing contrast. However particular care must be taken if colour is to be used in dialogues since an excessive use of colour can result in extremely confusing displays, and text in a variety of colours can be difficult to read. It must also be borne in mind that about 8% of the male population is colour-blind and cannot distinguish red from green.

Design of Validation Procedures and Error Messages

Validation rules must be established for each part of the dialogue that is input by the user.

Two points should be remembered when defining validation rules:

— To have an entry continually rejected can become extremely irritating. It takes time and can be frustrating, especially if the system gives no indication of how to correct the error.

One way of reducing this problem is to design a flexible system which can be programmed to allow the computer to reply 'DO YOU MEAN . . . . . ?' when common errors are made (Figure 4.10).

```
┌─────────────────────────────────────────────────────────┐
│ DATE                           4.11.73                   │
│                                                          │
│ DO YOU MEAN 04/11/73?                                    │
│                                                          │
│                                CUSTOMER ADDRESS?         │
│                                                          │
│ DO YOU MEAN CUSTOMER ADDRESS?                            │
└─────────────────────────────────────────────────────────┘
```

NB UNDERLINED = INPUT

Figure 4.10 Two Examples of Flexible Checking Systems

— A system with many validity checks can give the impression of a rigorous checking system that cannot accept invalid data. In fact, this can produce a false sense of security as illustrated in the example shown in Figure 4.11.

DATE?	
	16-4-80
INVALID ENTRY	
	16-04-80
INVALID ENTRY	
	16 APR 1980
INVALID ENTRY	
	04-16-80
INVALID ENTRY	
	16/4/80
FILE REQUIRED?	

Finally the input is accepted when in fact the correct date is 17 April 1980!

Because such rigorous checks are performed a false impression is created that the computer is infallible.

NB UNDERLINED = INPUT

Figure 4.11 Checking System Which Introduces a False Sense of Security

The effects of errors can be minimised if feedback is provided as early as possible. The aim of error messages should be to:

indicate that there is an error;

locate the error;

tell the user what to do next.

Error messages should be as detailed as necessary but no more than that, ie they should be explanatory and informative for inexperienced users, and brief and to-the-point for experienced users. There are three main ways of achieving acceptable error messages suitable in systems used by all levels of users:

— short coded error messages which can be expanded by pressing a 'help' or '?' key;

— lengthy, helpful error messages which can be interrupted by the user once they have reached the degree of help required;

— user performance monitored by the system which then selects the level of error messages suitable for the user in question.

Under no circumstances must the user be exposed to inexplicable error messages from the operating system.

Once the user has been notified of an error, he must be provided with the means of correcting it. Obviously the technique adopted will depend upon the dialogue style. For instance, with a menu technique the menu can be displayed and the request for input can be repeated. With a form-filling approach, correction by means of the cursor may be appropriate. Do not make the user re-input the whole screen as the result of making a single error. It may also be helpful in some circumstances if the validation criterion used to detect an error is shown on the same screen as the information that may have to be changed.

Formatting the Screen

The fundamental requirement for formatting displays is that the information on the screen should be organised and structured in a form appropriate to the activity and to the frame of reference of the user. Where several users may wish to perform several tasks with essentially the same information content, it may be necessary to compromise a particular user. However, the way in which information is organised and presented may determine how the material is used or the conclusions that are drawn from it. Simply providing users with the 'right' information and disregarding its presentation may prove to be false economy. Badly or inappropriately-formatted displays waste time and cause errors both in reading and interpretation.

Factors which contribute to a good format design are:

— logical sequencing;

— spaciousness;

— relevance;

— consistency;

— grouping;

— simplicity.

Logical sequencing. The sequence in which information is presented should be logical in terms of the system itself, the activity being undertaken and the other information sources (such as input documents) being used. Where these requirements conflict, it is the sequence in which the user requires the information which is of paramount importance, even though this sequence may not be logical to the computer or to the system designer. When it is not possible to suit the logic of the user and the activity he is carrying out, it is important that the user knows why the dialogue is sequenced the way it is.

Spaciousness. Spacing and blanks in a display are important both to emphasise and maintain the logical sequencing and structure of a dialogue and also to aid identification and recognition of items of information. Clutter on a display greatly increases search time and increases the likelihood of missing, overlooking or misreading items.

Relevance. One cause of clutter is the natural desire of the designer to ensure that the user has available all information of real and potential relevance to his task. In most situations it is better to exclude information of only 'potential' relevance from the primary display in order to ensure that the essential relevant information can be easily and accurately identified and read.

Consistency. The overall need for consistency in the language used in dialogue design has already been discussed. The formats also require consistency, both within and between particular dialogues. The value of consistency is that an unfamiliar or new output can be more readily and accurately interpreted.

Grouping. Where there are relationships between items of data and information the display can be improved if relevant items are grouped together. There is also evidence that displays of many similar items can be more rapidly and accurately searched if the items are grouped into manageable 'chunks'.

Simplicity. All the above factors should be taken into account in designing formats, but the overriding consideration should be to present the appropriate quantity and level of information in the simplest way. This does not mean that there is no place for highly detailed, complex displays; but even these can still be organised and structured, and should avoid unnecessary complexity. It is sometimes possible to overcome the problem by providing several simple displays rather than one complex one.

Testing and Modification

After the dialogue has been designed it should be tested in as realistic a way as possible, by the users, on the terminal equipment they will eventually be using. To ensure an effective test, a list of specific features to be tested should be drawn up beforehand. The test should then be structured to obtain data about each feature. This can be done by such techniques as automatically collecting error rates or input times, manual checking of the recorded test afterwards, or by asking users for their comments on specific features. The results of the tests should then be analysed and changes made to the design accordingly. This process should be continued until a satisfactory design is obtained.

Designing Flexibility into the Dialogue

Once a dialogue has been designed and tested for users of intermediate capability, it can be adapted to meet the particular needs of any other category of user. Inexperienced users will need more specific instructions and explanations. Very experienced users are likely to be aware of the sequence of events; the dialogue for them can be much abbreviated. Care should be taken to make it easy to pass from one dialogue level to the next, and to ensure that all users know how to change levels. All levels of dialogues should undergo the testing and modification stage described in Testing and Modification above.

DESIGNING PROCEDURES FOR FAULTS AND BREAKDOWNS

Introduction

Users may encounter faults or breakdowns in the day-to-day use of

— their terminals;

— their systems.

Faults merely reduce the efficiency of systems or terminals. *Breakdowns*, however, make system or terminals inoperable. Procedures must be designed to cater for both faults and breakdowns so that:

— the problem is dealt with and the fault or breakdown rectified;

— the user is able to take action when necessary to ensure the efficient continuity of the business operation;

— the user feels in control of the system and terminal by being able to take positive action in response to faults or breakdowns;

— the problem is recorded for the long-term analysis of fault and breakdown statistics.

Rectifying the problem For problems to be dealt with as quickly as possible, users should have a clear understanding of whom to inform about breakdowns or faults.

Ensuring continuity for the business operation To safeguard the business operation it is imperative to make arrangements for possible breakdown. Users must be capable of implementing these arrangements. This can be achieved by training the users and by documenting the arrangements. Trials of these breakdown procedures are sometimes held during the life of the system for auditing purposes. These help users to maintain their awareness of the arrangements.

User control in the face of faults and breakdowns When a fault or breakdown occurs, the user should always be able to take positive action. To avoid frustration, and lack of confidence in the system, every opportunity should be taken to make the user feel that he is in control of the system and equipment. This can be done by:

— providing feedback to users about the faults or breakdowns they have reported;

— giving users some responsibility in progressing the action taken over the faults and breakdowns they have reported;

— giving users the responsibility of posting notices at the workstation, informing other users of fault or breakdown conditions that have not been rectified (and of removing the notices, when the problems have been rectified);

— providing users with long-term fault and breakdown statistics.

Recording of faults or breakdowns A long-term record of faults and breakdowns is required so that:

— decisions about preventative maintenance can be taken;

— decisions can be made about the purchase or production of replacements.

The completeness and accuracy of fault and breakdown records depends almost wholly on the users logging all such problems. Users cannot be relied upon to do this merely by asking or telling them to do so. However, they will be more likely to do so if the following conditions exist:

— if they are confident that something will be done if they report or log the fault. Users should be told what has been done, in order to reinforce this confidence;

— if faults are dealt with quickly, and breakdowns are dealt with immediately.

5   Workplace Design

COLLECTION OF ENVIRONMENTAL DATA

Introduction

This section describes how to make a preliminary survey of a workplace in which new equipment is to be accommodated. The following factors should be covered:

— workstation design;

— lighting;

— room climate;

— noise;

— room layout.

The survey should demonstrate whether more detailed consideration needs to be given to any or all of these factors. If this is the case, the following sections provide relevant information.

The Preliminary Survey

The following data should be collected in the course of the preliminary workplace survey:

— workplace dimensions;

— layout of furniture and equipment;

— access requirements;

— floor characteristics;

— position of air conditioning or heating equipment;

- position and type of lighting;
- position and type of plug sockets and cable runs;
- noise and vibration characteristics of equipment.

Workplace dimensions

Workplace dimensions should be established first. Measurements should be taken to establish:

- room dimensions including ceiling height;
- position and size of windows;
- position and size of doors.

Two people will be required for this task. Room occupants should be told by their management what is to happen, and why, before the measurements are taken. When the measurements have been obtained they should be used to draw up a small-scale plan of the workplace on an A4 sheet of paper. Copies of this plan can then be used to record the remaining data that is required.

Layout of Furniture and Equipment

The layout of the room and its contents should be measured and detailed. Particular note should be taken of fixed, as opposed to freestanding, contents.

Access Requirements

Details should be noted of the following access requirements:

- direction of opening of doors and windows;
- free space required to open doors and windows fully;
- access to open and close windows;
- passageways within or through the workplace;
- frequency and route of regular movements of people or things through the workplace;
- access to, and use of, fire equipment.

Floor Characteristics

The strength of the workplace floor should be established and the nature of the floor covering noted.

Position of Air Conditioning or Heating Equipment

The position of radiators, extractor fans or any other air conditioning or heating inlets or outlets should be established, and if possible, ranges of heat outputs, air flows, and humidity. Also, it is essential to obtain the heat outputs of all equipment to be used at the workstation. Some equipment can produce so much heat that cooling of the environment will be required rather than heating, for a large part of the year.

Position and Type of Lighting

The position and type of all artificial light sources in the workplace should be noted, and if possible, illuminance levels and characteristics.

Position and Type of Plug Sockets and Cable Runs

Details should be noted of the following:

— the position and type of power and communication points;

— the route that any cabling takes from points to equipment;

— the state of repair of points, plugs and cabling.

Noise and Vibration Characteristics of Equipment

Information should be obtained from manufacturers relating to the noise emission and vibration characteristics of all equipment.

WORKSTATION DESIGN

Introduction

A workstation is a position in a workshop at which work is undertaken by a human being with the aid of physical equipment and facilities. The objective of workstation design is to design, select and arrange this equipment and these facilities to create a physical framework which promotes efficient operation and personnel well-being and safety. Particularly undesirable features, which ought to be regarded as hazards, and eliminated where possible are:

— postures which cause muscular fatigue;

— visual environments and tasks which cause visual fatigue;

— other health or safety hazards (eg sharp edges on equipment).

Muscular Fatigue

Dynamic muscular work involves alternate contraction and relaxation of muscle tissue, which increases the rate of blood flow through the muscle. The blood vessels in the muscles are equipped with a system of valves. When the muscle is contracted it becomes hard and the blood vessels are drained; when the muscle is relaxed, the blood vessels are filled again. This pumping mechanism, which is called a venous or muscular pump, is very effective. By means of this blood flow, during dynamic work the muscles have ready access to energy-releasing material and can remove waste products which impair muscle function. Static muscular work involves continuous contraction of muscle tissue and the conditions for blood circulation are quite different. The muscular pump stops functioning, and it becomes more difficult to supply blood because the contracted muscle becomes solid and hard; the muscle receives very little blood and quickly tires. The greater the load, the less blood the muscle will receive and the quicker it will tire. At maximum load the muscle becomes completely fatigued after only a few seconds. If the load on a particular muscle being exercised dynamically amounts to no more than 15% of that muscle's capacity, it will function satisfactorily for an unlimited time.

Visual Fatigue

Visual fatigue can occur in any task, but work involving specific visual tasks, such as using a VDU screen is more likely to cause this problem.

Visual fatigue is a general term, used more to describe a variety of symptoms than a particular condition of the eye and its associated parts (such as the visual cortex — that part of the brain related to sight). The physiological causes of these symptoms are numerous. For example, the small muscles which control the focal length of the lens in the eye can become fatigued, causing the sufferer a sense of pain in the eye, behind the eye, or just a 'headache'. Postural strains on the muscles of the neck and face can also cause headaches, and are related to visual fatigue. Other symptoms are itching eyes, spots before the eyes, and difficulty in changing focus from near to distant viewing.

Potential task- and equipment-related causes of visual fatigue include the following:

— constant re-focussing of the eye to accommodate differing viewing distances;

— frequent eye movements near the extreme ranges of available movement;

— continuous focussing at a near point for prolonged periods (more than 30 minutes);

— undesirable aspects of the task and equipment causing postural strain and fatigue;

— undesirable aspects of the visual environment, eg excessive glare from over-bright light sources, insufficient levels of illuminance at the workstation;

— eye defects which are wholly or partially uncorrected.

Appropriate workstation and task design can eliminate or reduce the effects of the first four causes. The fifth cause can be greatly reduced by applying the principles enumerated later in this section. The sixth cause, which relates most personally to the worker, requires ophthalmic assistance. In most cases this will merely involve the usual NHS eye test, and corrective lenses in the form of spectacles or contact lenses will be prescribed.

Other Health and Safety Hazards

This category contains all the other health and safety hazards in the work-station, most of them relating to the equipment provided, and might involve:

— Being struck by	— sharp edges;
	— pointed corners;
	— falling equipment which was unstable;
	— moving equipment or machinery;
	— moving parts of equipment or machinery (eg some photocopiers have moving parts which could hit the face/eyes of the person leaning over the table on which the machine is placed).
— Being injured handling objects	— which are too heavy or cumbersone to be handled in the way so done;
	— which are too heavy or cumbersome to be handled in the way so done;
— Environmental effects	— toxic or noxious fumes given off by equipment, during normal operation;
	— toxic or noxious fumes given off by equipment during breakdown;
	— fire caused by equipment faults or failure.
— Electrical shock	— from 'faulty' equipment;
	— caused by inappropriate action by personnel.

Most of these hazards can be eliminated by appropriate selection of equipment, and proper checking on delivery for faults, also by clearly stating appropriate and inappropriate personnel actions and activities, and reinforcing these during training, and perhaps by other aide-memoires on the job, eg 'under no circumstances remove this panel until all power to the device has been switched off'.

Design Parameters

To undertake the design of a workstation, details of the following are required:

- equipment to be used;
- activities to be undertaken in conjunction with the equipment;
- other activities to be undertaken at the workstation;
- people who will be using the workstation;
- the workplace in which the workstation is to be located;
- workstations currently in use.

Equipment

The following aspects of terminal equipment may place constraints on the workstation design:

- *the setting and monitoring of controls.* The operations and information displays that allow control over the equipment must be listed and an assessment made of their associated postural requirements and limits.
- *use of the equipment.* A detailed breakdown of the actions that must be undertaken by specific parts of the body to use the equipment is required. Then, using this breakdown, an assessment must be made of the associated postural requirements and limits.
- *maintenance of the equipment.* Likely maintenance activities must be identified and an assessment made of the access requirements to support those activities.
- *the installation or replacement of the equipment.* An assessment must be made of how the physical installation or removal of the equipment will be achieved to ensure that workstation design does not impede the activity.
- *the services which must run to and from the equipment.* Terminal equipment usually has two service connections — power and communications. However, whatever service links are required, they must be identified and provision made for them.
- *the size, weight and stability of the equipment.* These features all determine the size and strength of the surface on which the equipment will sit.

An example of how these six aspects were assessed for a visual display unit and keyboard are shown in Appendix 4.

Activities to be Undertaken in Conjunction with the Equipment

The activities to be undertaken in conjunction with the equipment and the methods by which they are carried out should have already been established at the job design stage. The following aspects of each activity should be identified:

— use of documents or reference manuals;

— interaction with other people;

— use of other equipment such as telephone or calculator;

— any writing that is undertaken;

— the duration of the activity;

— the frequency of the task.

Other Activities Undertaken at the Workstation

Some activities that do not require the operation of the terminal may also have to be carried out at the workstation. These too should have been identified at the job design stage and should be considered in the same way as the tasks above.

Users

The workstation must be designed to comply with the constraints imposed by the range of physical dimensions and capabilities of the potential users. The potential users consist of existing and future personnel. To establish the existing population, reference should be made to the list of users established at the identification stage. This information should then be used to answer the following questions:

Can the users be personally identified? If so, who are they?

— male/female?

— handicapped? If so, state handicap.

— age range.

— physical characteristics, height, etc.

If the users cannot be personally identified, attempt estimates or predictions of the following:

— male only/female only/both.

— age or age range.

- handicaps?

- physical characteristics.

In the case of future users the important factor to determine is whether there is any intended or proposed change in user population. This might be caused by imminent legislation, or even company policy changes; for example, the inclusion of women in a previously all-male domain, or the inclusion of certain types of disabled people who hitherto would not have been considered for this work. Having established if there is to be a change, its implications for workstation design have to be spelt out, eg wheelchair users must be capable of operating all the equipment at this workstation (and others). This would affect space and access requirements of personnel, required ranges of adjustment of equipment, etc.

The Workplace in Which the Workstation is to be Located

The size of the workstation may be limited by the space available in the room. The nature of the floor will determine the type of base that the workstation will sit on, and the strength of the floor will limit the maximum loaded weight of the workstation.

The Workstations Currently in Use

This data may provide insight into special requirements or things to avoid in the design of the new workstations. It also allows current workstation suitability to be assessed against the new design when it has been completed. To obtain this data, the workstations should be measured up and notes made of the facilities they provide and the general condition they are in. Current users should be asked for their opinions of the workstations and encouraged to point out their failings and good points. Ensure that seating is included in these investigations.

Workstation Design

The workstation design process can be split into 6 stages:

- establishing initial design parameters;

- establishing worksurface shape, dimensions and layout;

- chair and footrest design;

- detailed furniture design;

- checking the design;

- testing the design.

Before embarking on the workstation design process, the following 3 maxims should be remembered:

- the design process must take place using anthropometrical and psychological data for a specified user population (eg male and female adults between the ages of 16 and 65 and not having any physical handicap).

- as the design is built up, conflicting requirements will almost certainly be encountered. Successful workstation design depends on making optimum compromises.

- as much flexibility as possible must be built into the design to cater for variation in the existing user population, possible future user populations (eg including certain types of handicapped people), and probable future users of the equipment.

Establishing Initial Design Parameters

Six basic design parameters can be investigated initially:

- defining the user population;
- sitting or standing workstations;
- heights;
- viewing distances;
- eye and head movements;
- document holders.

Decisions about each of these parameters should be taken at this initial stage. The remainder of the design can then be built around these parameters. If compromises have to be made later in the design process, these parameters can be modified accordingly at that point.

Defining the User Population

For a workstation to promote efficient operation whilst safeguarding the health and safety of personnel using it, its design must take due account of the physical dimensions and capabilities of the personnel. Anthropometric data will be required for this purpose. (Anthropometry — which literally means measurement of man — is a scientific discipline concerned with the measurement of the physical dimensions of the human body.) Also associated with this is the measurement of the ranges of movement available at the various joints, and the forces and degree of control exerted by numerous muscle groups.

In the case of physical dimensions, any measured population is usually broken down into percentiles, or 1/100ths. The 1 percentile would consist of

the first hundredth of the population, the 5 percentile the first five hundr-
edths, etc. These quantities are statistically derived, and do not necessarily
represent any one particular individual. In the design of workstations, the
most critical percentiles are usually the 2.5 percentile and the 97.5 percentile.
These are used as expressions of the extremes of the range of users' dimensions which means that any such design will fit the middle 95% of the population. Following Equal Opportunities legislation, it is now usual to specify the
percentiles for both male and female. This gives a wider range of values since
the 2.5 percentile for females in most dimensions is smaller than the male and
the 97.5 percentile for males is larger than the female.

If the workstation is to be designed for one person, or a group of known
persons, the usual anthropometric data should still be used, but additionally, if
the known persons are outside the 2.5 or 97.5 percentiles, make sure that
they can be fitted into the workstation by appropriate changes to equipment
adjustability, available spaces, etc. Anthropometric data is used in this case to
prevent the workstation being so individually tailored that when the person
concerned left the task, it would not fit any new recruit! Since you cannot
accurately predict the physical size of any new recruit, you must cater for
95% of the population, again.

Sitting or Standing Workstations

Working whilst standing is undesirable in most cases, since it involves static
loading of a considerable number of muscles to maintain an erect posture and
these muscles can soon become fatigued. Standing absolutely still is more fatiguing since the muscles are even more tensed. The necessity to lean out of
the upright position (eg slight forward), for long periods requires static operation of muscles in the back, causing rapid fatigue and, often, backache.
Wherever possible, therefore, standing for long periods is to be avoided. If,
however, the nature of the task, or the equipment layout necessitates a standing work position, attention should be paid to the points raised, so that the
person can stand in a natural posture, and be able to change position and
posture, if only slightly. This change of posture allows different groups of
muscles to be used in slightly different ways, alleviating the build-up of fatigue.

Most office work is carried out in a sitting position and this should reduce
muscular fatigue, stabilise the location of the occupant, and provide support
at the buttocks and lumbar region of the spine. However, the seat and the
work space should permit a variety of postures to be adopted. This again
allows different groups of muscles to be used in different ways in the various
postures, offsetting fatigue caused by static loading. Critical surfaces determining good working postures are the floor level, seat cushion height and the
height of the workplace or work.

Heights

The correct height for a chair seat is three centimetres less than the distance between the hollow of a person's knee and the floor, with the knee bent at a 90° angle. (The height of the shoe heel is included in this measurement.) The entire sole of the shoe should then rest on the floor, whilst the seat does not exert too much pressure on the back of the thigh.

The thigh muscles should be relaxed when the measurement is taken; otherwise the tendons in the hollow of the knee will be forced downward and the measurement will be incorrect. This is because the flow of blood to the legs is impeded if pressure is applied to the blood vessels in the lower part of the thigh. Subtracting three centimetres ensures that the thigh will not exert too much pressure against the front edge of the seat.

A fixed chair height of 43 cm will accommodate the largest number of male adults but a footstool, adjustable up to 5 cms, may be necessary for the small female. Adjustable chairs should be variable in height between about 34 and 52 cm.

Worksurface heights for seated operators are dependent on:

— *thigh clearance requirements*. The minimum amount of space provided for thigh clearance should be 18 cm measured from the front edge surface of the seat to the underside of the worksurface. However, in practice people do not sit with their legs still all the time. In order to remain comfortable in the seated position some amount of movement and fidgeting is necessary and many people like to sit for varying lengths of time with their legs crossed. Therefore it is advisable to provide as much thigh clearance as possible.

— *the activities to be undertaken at the worksurface*. Manual operations are performed best when the worksurface is either level with or below the elbows and the angle formed at the elbow is 90° or more. Using this maxim the optimum height for normal office work — writing, reading, and manipulation of paper — is 71 cm for males and 68 cm for females for a fixed surface or 67—79 cm for an adjustable surface. For typing activities, however, the height of the keyboard must be taken into account. So in this case, a lower worksurface height of 64 cm for males and 61 cm for females for a fixed surface and 52—67 cm for an adjustable surface is recommended.

If a number of different activities are to be carried out at the same workstation, a number of surfaces may have to be provided, each at the appropriate height.

— *worksurface thickness*. This effectively reduces the amount of thigh clearance space. Therefore, as a general rule the worksurface thickness should be kept to 2 cm or less.

Clearly the amount of time and effort to be spent on workstation design and the novelty of the application for which it is required will dictate the degree to which heights will be considered. At the one extreme, tables of anthropometric measurements will be required or even anthropometric measurements taken on the appropriate population. At the other extreme, the recommended heights shown in Figure 5.1 can be used.

	Fixed Female	Fixed Male	Adjustable
Chair seat	41 cm	43 cm	34 − 52 cm
Worksurface			
— writing	69 cm	71 cm	67 − 79 cm
— keying	61 cm	64 cm	52 − 67 cm

Figure 5.1 Recommended Chair and Worksurface Heights

It is important to build as much height flexibility into the workstation as possible. However, such flexibility will be wasted unless users can actually adjust the equipment to suit themselves. Unfortunately many users are either not aware of the parameters which affect seat height, or find the necessary adjustments too difficult and time-consuming. Therefore for workstation design work to be relevant to the real world, it will be necessary to include information about correct seating postures, worksurface heights and how to adjust the console in user training and to ensure the adjustments are easy to make.

Viewing Distances

Maximum viewing distances for displays are generally accepted to be 71 and 33 cm respectively with an optimum distance in the region of 45–50 cm. Therefore the workstation design should try to achieve viewing distances as close to the optimum as possible and certainly well within the maximum and minimum range.

Optimum viewing distances begin to change with age from about the late 30s onwards: the focal length becomes longer. However, unless there is specifically an older population to cater for, the design should proceed using the distances already recommended. The needs of older users can best be dealt with by the use of spectacles or contact lenses.

The eye is susceptible to strain if it has to refocus continually over a period of time. Therefore, perhaps it is more important to ensure that the distances between the eye and the viewing surface, keyboard and source/working documents are kept as similar as possible as shown in Figure 5.2, than to achieve optimum viewing distances.

However, when the eye has to maintain one focal length continuously for long periods, the intrinsic muscles controlling the focal length of the eye become fatigued. When a person views distant objects (25'+), the intrinsic muscles are able to relax, since this focal length is provided by the lens in its natural shape. If possible, a distant view (25'+) should therefore be available from the workstation. Periodic distant glances can then be made, offsetting visual fatigue.

Eye and Head Movements

Maximum and optimum eye movements and the range of easy head movement are shown in Figure 5.3.

While it is undesirable to design a workstation which necessitates a lot of head movement in order to use a terminal, nevertheless eye movements of an optimum or slightly greater level can be supplemented *occasionally* by small head movements. Such head movements are likely to be relaxing and beneficial to the operator.

Document Holders

If source documents are placed flat on a desk next to a keyboard, undesirable head or body movements will be required to read them. Therefore, where possible, a document holder should be used. The following points should be taken into account when designing document holders:

— head movements between the screen and the source document should be restricted to the horizontal plane;

— if a source document is difficult to read, the user will probably try to reduce the viewing distance by changing his posture. Such a posture change is easier to effect if the document is next to and in the same plane as the screen, than if it is lying flat on the desk next to the keyboard;

— the most comfortable head and eye positions when reading in a seated position are usually with the head declined at approximately 20° and with the eyes declined at approximately a further 20°;

— if the source document is lying flat on a desk the luminance ratio between screen and source document will be excessive — in the region of 1:6. If the source document is inclined at an angle of 20° to the vertical, the luminance ratio is likely to be more acceptable — in the region of 1:3;

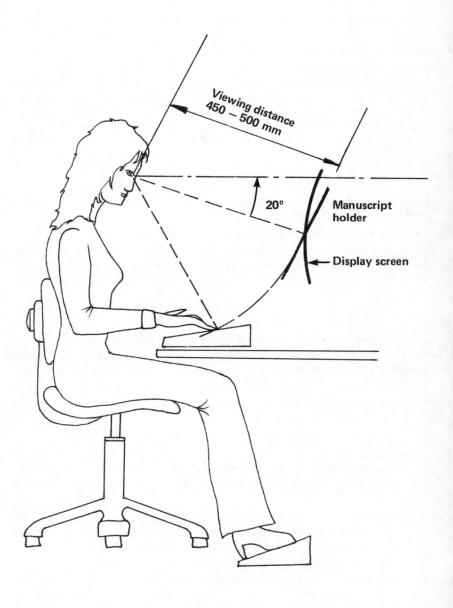

Viewing distance
450 – 500 mm

20°

Manuscript
holder

Display screen

Figure 5.2   Recommended Viewing Distances

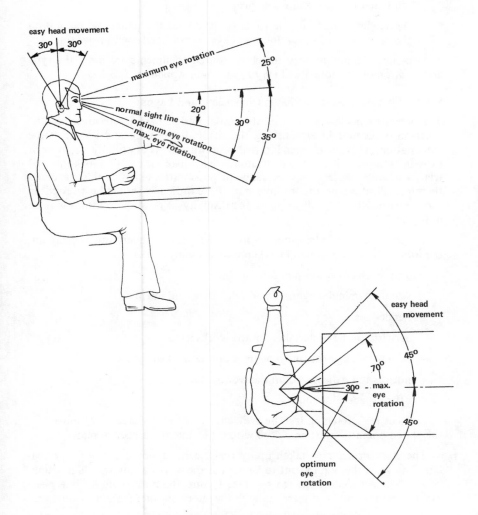

Figure 5.3   Maximum and Optimum Eye Movements
and Range of Easy Head Movements

- some flexibility should be built into the design of the document holder so that individuals can make slight adjustments to its height and inclination and to the reading distance;

- the reader should be able to move the document holder and place it on the same side as his/her dominant eye and/or hand used for writing.

Depending upon the way in which source documents are used, the type and position of the holder will vary as shown in Appendix 5.

Establishing Work Surface Shape, Dimensions and Layout

The decisions made for each of the initial design parameters must now be taken into account in establishing the shape, dimensions and layout of the workstation. There is no standardised procedure: it is essentially a matter of experimentation with a range of alternatives, making use of mock-ups and trials wherever feasible. A wide range of possibilities exists at this stage and therefore all ideas, no matter how out of the ordinary, should be assessed on their own merits. An indication of the diversity of possible solutions is given in Figure 5.4.

In the course of establishing worksurface shape, dimensions and layout, the following factors should be taken into account:

- arm reach (forwards and sideways);

- size of terminal, keyboard and any other equipment;

- handedness;

- activities to be undertaken on the worksurface;

- documents or manuals to be used on the worksurface;

- notices to be displayed on the worksurface;

- office space constraints.

If excessive stretching is to be avoided, items to be picked up or moved by the seated operator must be placed within the operator's normal reach.

The worksurface area taken up by the screen, keyboard, telephone, calculator and any other equipment to be used at the workstation, should be established. When designing the worksurface layout, the *positioning* of one piece of equipment should not interfere with the *use* of another piece of equipment.

Workstations should cater for both right- and left-handed people. This point should be borne in mind when considering the activities that are to be carried out at the workstation.

Attention should be paid to the equipment and worksurfaces that will be required in the course of each activity with respect to activity content and

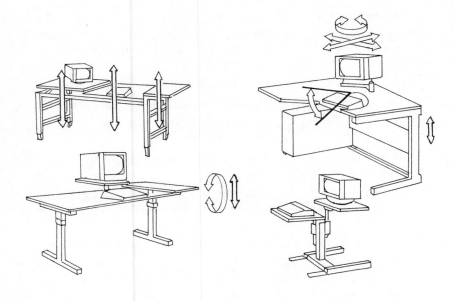

Figure 5.4   An Indication of the Diversity of Possible
Workstation Designs for VDUs and Keyboards

frequency. This knowledge should assist in the design of worksurface layout.

The use of documents and manuals on the worksurface is likely to constitute the largest space requirement apart from the equipment itself. The following requirements should be particularly noted;

— the need to read two or more documents or manuals in conjunction with one another;

— the sorting of documents;

— the need to make piles of documents to keep control of them;

— the need to write on the documents.

There is often a need to display notices, eg equipment operating instructions, systems operating instructions, notification of temporary faults or changes, terminal identification, on workstations.

Whatever the need, there are four criteria for notice display:

— the notice must be not only visible and readable but also eye-catching;

— the notice must not obstruct the activities undertaken at the workstation;

— if the contents of a notice are required during an activity, it must be readable during the task;

— the notice must be durable and should not be positioned where it may become illegible through being torn, being screwed up or becoming dirty.

The space available in an office may constrain the size of workstation.

Items of equipment, displays, controls etc, can be laid out according to one of several methods. Firstly, by importance — where those controls, items of equipment judged to be the most important to the task at the workstation occupy the prime positions. Secondly, by function — where items of equipment, control etc, related to one function are grouped together and segregated in one area of the workstation. Thirdly, by frequency of use — those items of equipment, controls etc, used most frequently are located nearest to the operator; those used only occasionally will therefore appear on the periphery of the workstation. Fourthly, the items of equipment, controls etc, can be grouped according to sequence of use. There are often sequences or patterns of operation that occur frequently, and this sort of arrangement takes advantage of one of these sequences. As a general rule such grouping should provide operator movements from left to right and top to bottom. This type of grouping is often an aid in reducing omissions of operation.

Chair and Footrest Design

Chairs for terminal equipment workplaces should be adjustable in height between 34 and 52 cm. The height adjusting mechanism must be simple, easy and safe to operate from the seated position. This is particularly important if the chair is to be used by more than one person on a regular basis. It may also be appropriate for users to place their own height adjusting marks on these chairs if it will simplify the adjustment procedure.

The seat should be horizontal. It can be slightly concave, but it should not be designed to exactly fit the body, since this makes turning to the side difficult. Thin upholstery, foamed plastic for example, is suitable. Generally, thick upholstery is too warm. A seat size of about 40 x 40 cm is recommended.

As a rule, chairs should be provided with backrests. It is not actually the back that is primarily in need of support but the pelvis. The pelvis can move forward and backwards, and is held in position by muscles. If the pelvis is supported at the small of the back, the muscle load can be reduced. Therefore an adjustable backrest should be positioned at the back of the waist. In practice backrests are often set too high. High backrests may be valuable as a support when one is sitting in a resting position, but in many cases, they tend to hinder movement.

For normal office work, chairs with armrests should be provided. If the chair is to be used at a VDU workstation, armrests should be avoided as they interfere with efficient keyboard operation. However, if the work involves a considerable amount of data entry using a numeric keypad, the support provided by armrests can be helpful to the operator.

A swivel facility can be advantageous since it extends the lateral reach and facilitates getting in and out of workstations.

Chairs with casters can be moved about easily. However, keeping such chairs in position often requires static muscular effort. The combination of type of caster and type of floor covering will determine the suitability of such a chair. If a typist chair is chosen it should have a five-point base to ensure stability.

Footrests should be provided at workstations to ensure that smaller people can adopt the correct leg posture, ie the front of the thighs slightly raised off the front surface of the chair. Footrests should be height-adjustable in the range 0—5 cm. It should also be possible to adjust their angle of inclination in the range from 10 to 15 degrees. The footrest base should not allow the footrest to slide around the floor when in use.

Detailed Furniture Design

Although basic dimensions of the chair and workstation, and workstation lay-out are of prime importance, other minor design details will play a critical role in workstation acceptability and usability.

The following are of particular importance:

- leg space;
- storage space;
- modularity;
- ease of cleaning;
- cabling;
- materials;
- weight;
- mobility;
- floor covering;
- waste disposal facilities;
- safety.

A minimum of 80 cm width and 70 cm depth is required to provide adequate leg space, and the leg space area should not be obstructed by desk frame spars, etc. It should not be possible for the legs to encounter any sharp edges or pointed corners in position at the workstation, or when getting in or out of the workstation.

All requirements for storage space at the workstation should be identified. Most common requirements are for user's guides to the system and for personal effects (ladies' handbags in particular). If the workstation can accommodate more storage than is needed, this could be an ideal opportunity to take care of other unrelated storage needs in the office.

When all storage requirements have been established, the detailed storage space design can be undertaken. The following points should be taken into account:

- access to all items stored must be as easy as possible;
- locks may be required.

A modular design will enable two or more workstations to be positioned together. There are three main reasons why a degree of modularity may be useful:

- when office space is limited, modularity may allow for more furniture to be accommodated in less space;
- the grouping of workstations that modularity facilitates enables central power and communication points to be used, thereby reducing cable runs across the office;
- modularity facilitates the setting up of small working groups.

Dust and dirt will collect on worksurfaces and in nooks and crannies such as certain types of drawer handles. The design should exclude as many of these potential dust and dirt collectors as possible. All such parts that appear in the final design should be listed and an assessment made of how cleaning can be achieved. Ease of access to the open floorspace underneath the workstation for cleaning purposes should also be catered for.

Loose cabling underneath desks, on top of desks or dangling off the edge of desks can obstruct people. If it gets pulled accidentally it may even cause injury, damage to the connections or the pulling of the equipment off the workstation onto the floor.

Cabling also makes an office look untidy, collects dirt and dust, and can obstruct office cleaning. Possible ways of combating some of these problems include running the cable through specially designed channels in the workstation or attaching cables to the outside of the workstation with adhesive cable clips. Solutions will vary with different applications. The important thing is to ensure that cables are out of the way and secure.

The material from which the workstation is to be constructed should:

- not induce reflected glare. The worksurface should be matt finished with a reflection factor of 0.4, or 0.6 as a maximum;
- not be abrasive or splintering;
- be easy to clean;
- be hardwearing and durable.

If writing is to be undertaken, the worksurface material must be firm and even.

The surface on which the workstation is to stand may constrain the weight of the loaded workstation. Workstation weight will affect mobility. As flexibility is often an important requirement in offices, workstations must be easy to move around. The easiest way of achieving mobility is to fit braked wheels to the workstation. These should be tested for stability in the braked condition before being used.

The choice of floor covering material affects a number of factors:

— noise levels will be low if an absorbent material such as carpet is used, and higher if materials such as lino tiles are used;

— static electricity is produced by friction when people move about on carpets made of synthetic material. The static charge then builds up and small shocks can be received from metal furniture etc. Although such shocks are not dangerous they can be disconcerting. To eliminate static, special antistatic carpet can be obtained, or a chemical solution can be sprayed onto carpets at approximately 6-monthly intervals;

— desks must remain stable when standing on the floor surface. If desks are on braked wheels, the floor surface must allow movement of the desk when required;

— chairs must be moved to sit down at and get up from the workstation. In addition, small adjustments to the chair position are often made while the user is seated at the workstation. The floor surface must allow the user to achieve these chair movements with little effort, but not so easy that control can be lost. This latter case arises when chairs with free-running casters are placed on hard, smooth, lino flooring.

Adequate waste disposal facilities should be available at the workstation: these should be within easy reach, should not obstruct the users' legs and feet when seated at the workstation, and should not obstruct adjoining passageways or access to the workstation itself.

All exposed corners should be rounded. Adequate leg room should be provided for the user, and knees and legs should not come into contact with any sharp edges or corners. Cabling must not be located in a position where it might be damaged (by closing drawers for example), or tripped over.

Checking the Design

Once the workstation design work has been completed, a final check that the design allows the following activities to be undertaken should be carried out:

— the setting and monitoring of the terminal controls;

— use of the equipment;

— maintenance of the equipment;

— the installation and replacement of the equipment.

If the design stands up to this final check successfully, a prototype should then be produced and tested.

Testing the Design

No matter how carefully the design work has been undertaken, only realistic trials will give a true picture of the workstation's performance. A list of factors to be evaluated, and details of the evaluation procedure for each factor, should be drawn up before the trials. This list of factors will vary from application to application. However, a typical list for a VDU workstation is shown in Appendix 6.

Workstation trials should be undertaken bearing in mind the three types of hazard discussed previously:

— muscular fatigue;

— visual fatigue;

— other health and safety hazards.

By the end of the trials it should be known whether any of these hazards exist for the workstation users.

The design may have to be modified after the evaluation procedure, and tested again until the design is satisfactory.

LIGHTING

Basic Lighting Concepts

In any workstation there are times when the natural light is not adequate for the tasks(s) to be carried out there. This means that natural light must be augmented by artificial light from light fittings appropriately selected and fitted in the workstation. Different tasks require different levels of acuity (the ability to see fine detail), and the latter is very dependent on the light levels in the task area. Also the number, brightness and location of light fittings, and other factors in the workstation can cause glare, which may impair performance and cause discomfort. Therefore the two main aims in introducing artificial lighting to a workstation are to provide light levels appropriate for the tasks to be carried out efficiently, and to ensure the visual environment does not cause glare, or predispose personnel to visual fatigue.

In order to appreciate the lighting requirements of computer terminal equipment work, it is necessary to understand some basic lighting concepts.

The light arriving at the eye, which has come from a source via a reflection from a surface, can be considered in the following ways:

— the source;

— the flow;

— its arrival;

— its return.

This is illustrated in Figure 5.5.

Figure 5.5 Basic Lighting Concepts

— *The source*: The luminous intensity (I) describes the power of the source and is expressed in candelas (cd).

— *The flow*: This is described as the luminous flux and is measured by means of the lumen (lm).

— *The arrival*: The amount of light arriving at a surface is called the illuminance (e) and is measured in lux (lx). (1 lux = 1 lumen/square meter.)

— *The return*: The amount of light reflected from a surface is called the luminance (L), (or brightness) and is measured in candelas per square metre (cd/m^2) or apostolibs (lumens/square metre).

Assessing Lighting Requirements

The assessment of lighting requirements can be considered as a seven-stage process:

A Review of existing lighting scheme

B Identification of activities

C Establishment of recommended light levels

D Measurement of existing light levels

E Assessment of glare problems

F Assessment of reflection levels

G Review of new lighting scheme

A Review of Existing Lighting Scheme

The existing lighting scheme can be reviewed by:

— referring to the lighting information obtained during the collection of environmental data;

— asking the current room occupants their views on the existing lighting scheme.

B Identification of Activities

The purpose for which the room is used, and the various activities this involves, must be identified. The latter should have been identified at the analysis stage. The information thus generated will help to establish the recommended light levels and types of lighting for each activity. It will then be possible to determine if the various needs of the different tasks can be met by one lighting assembly, or whether a number of lighting levels and lighting assemblies are required.

C Establishment of Recommended Light Levels

Recommended light levels have been determined by the Illuminating Engineering Society (IES) and are documented in a handbook called 'The IES Code'. The recommended levels in the IES code are based on:

— the different likely characteristics of many defined visual tasks;

— practical experience in the lighting of those tasks;

— the need for cost-effective use of energy.

The light levels in the IES code are given in terms of the 'Standard Service Illuminance'. This is the mean illuminance of the lighting system throughout its maintenance cycle (ie from when lamps are installed to when they are replaced), and is averaged over the area in which the activity takes place. The recommended Standard Service Illuminance may need to be raised in circumstances where errors will have serious consequences or where there are no windows. If the task is of short duration, the recommended level may be lowered. The IES code provides guidance about these adjustments.

The code presumes that the lamps will be replaced regularly and that the walls and light fittings will be kept clean. Figure 5.6 shows extracts from the code covering a wide range of jobs. Figure 5.7 shows extracts of the code for office situations.

The activities that are to be carried out in the room concerned should be matched to their closest equivalent in the IES code and the recommended Standard Service Illuminance identified. If a variety of activities require different levels of illuminance, a general lighting scheme will be necessary with supplementary lighting being provided to supply the special illuminance requirements of the more visually exacting tasks.

The use of Visual Display Units had not been included in the 1977 edition of the IES code (the latest version available in 1980). However, the manual *Visual Display Terminals* recommends that for VDU operation, no more than 500 and no less than 300 lux illuminance should be provided as measured on the surface on which the VDU is standing.

	Standard Service Illuminance (Lux)
Storage areas and plant rooms with no continuous work	150
Casual work	200
Rough work (Rough machining and assembly)	300
Routine work (Offices, control rooms medium machining and assembly)	500
Demanding work (Deep plan, drawing or business machine offices. Inspection of medium machining)	750
Fine work (Colour discrimination, textile processing, fine machining and assembly)	1000
Very fine work (Hand engraving, inspection of fine machining or assembly)	1500
Minute work (Inspection of very fine assembly)	3000

Figure 5.6 Extract from the IES Code Showing a Range of Recommended Standard Service Illuminances

	Standard Service Illuminance (Lux)	Position of measurement
General offices with mainly clerical tasks and occasional typing	500	Desk
Deep plan general offices	750	Desk
Business machines, typing offices, punch card rooms	750	Copy
Filing rooms	300	File labels
Conference rooms	750	Tables
Executive offices	500	Desk
Banking halls:		
— working spaces	500	Desk
— public spaces	300	Floor
Computer rooms	500	Working Plane
Drawing offices:		
— drawing boards	750	Board
— reference tables and general	500	Table
— print rooms	300	Table

Figure 5.7 Extract from the IES Code Showing Standard Service Illuminances For Office Situations

It has been suggested that, in order to reduce screen reflections and to increase the relative luminance of the characters on the screen, less than 300 lux illuminance should be provided. This is not recommended. Screen reflections are best dealt with by appropriate positioning of the VDU, and adequate character luminance should be one of the criteria by which a VDU is selected for purchase.

The IES code also recommends lamp colour appearances that are suitable for specific task situations. The classification of lamp colour appearance is cool, intermediate or warm. Intermediate or warm lamps are preferred for working interiors since cool lamps tend to make a room look rather gloomy. When colour matching forms part of the activities to be undertaken in a room, special care must be taken to select lamps specifically recommended for such activities. When both artificial lighting and daylight are to be employed, fluorescent lamps of an intermediate colour appearance are the most satisfactory. Lamps of different colour appearances should not be mixed in a particular room. Figure 5.8 describes the colour appearance and typical applications of a selection of tubular fluorescent lamps. New lamps are always coming onto the market and manufacturers should be consulted for their complete ranges when required.

Lamp name	Colour Appearance	Typical Application
Artificial Daylight	Cool	Used for critical colour matching
Natural	Intermediate	Offices, department stores
Colour 84	Intermediate	Offices, department stores
Warm White	Warm	Restaurants, homes
Softtone 32	Warm	Restaurants, homes

Figure 5.8 Colour Appearances of Some Common Lamps

D Measurement of Existing Light Levels

Having established recommended light levels and appropriate lamp colour appearances, the existing light levels must be measured to determine whether they are satisfactory. The measurement of light levels is called photometry and is undertaken using a photometer. There are many different photometers on the market but a fairly simple and relatively inexpensive one can be used to measure illuminance levels. A typical photometer is shown in Figure 5.9.

To measure illuminance levels the photocell should be placed in the position recommended by the IES code for a specific task. The photocell should

Figure 5.9   A Typical Photometer

not be handled as fingerprints on the glass will affect the light levels recorded. The meter should be moved as far away from the photocell as possible so that the person reading the meter does not cast a shadow over the photocell; lux levels can then be read directly from the dial on the meter.

The existing light levels should then be compared to the recommended levels. If the existing lighting scheme is inadequate, a number of options are open to improve it:

— a completely new lighting scheme can be designed and installed;

— additional lamps can be fitted;

— the existing lamps can be upgraded to produce greater illuminance;

— local lighting can be installed to meet the needs of specific tasks.

If drastic changes to the existing lighting are being considered, it is advisable to call in a lighting engineer.

If fluorescent lighting is to be installed, the possibility of flicker should be reduced as far as possible. Flicker can be perceived from fluorescent tubes especially as they begin to age, the flicker being more noticeable at the ends of the tube. To avoid flickering at the end of a tube's life, special starters can be purchased which prevent the tube illuminating if it is going to flicker.

Flicker through the period of a tube's life can be prevented by using a three-phase mains supply and dividing the lamps between the phases.

E Assessment of Glare Problems

Glare occurs when there are highly contrasting illuminances in a person's field of vision, ie when bright sources of light such as lamps, windows or their reflected images are too bright compared with the general brightness of the interior. The physical effect of glare is to disturb the process of visual adaptation. There are two types of glare:

— disability glare or dazzle impairs the ability to see detail without necessarily causing visual discomfort;

— discomfort glare causes visual discomfort without necessarily impairing the ability to see detail.

Both types of glare can be caused either directly by the light sources (direct glare); or by the reflection of light sources from illuminated surfaces such as walls and desktops (reflected glare).

Glare increases with increasing luminance of the luminous surfaces and their size within the visual field, and decreases when the glare sources are more remote from the viewing direction or when the areas surrounding them are made brighter. Hence less glare is perceived from ceiling lights if the ceiling is brightly painted. Some people are more susceptible to glare than others, and therefore varying reactions to glare problems are to be expected.

The glare index is a series of numbers in the sequence 10, 13, 16, 19, 22, 25 and 28. For a given set of room and lighting conditions, the glare index can be calculated using the method described in the Illuminating Engineering Society Technical Report No 10.

However, it is a complex procedure and, if such measurements are required, a lighting engineer should be employed to undertake the work. In cases where glare is to be measured, the results can then be compared with the glare indices recommended in the IES code for specific activities. However, it should be possible to identify any likely glare problems by simply looking for them when in the room concerned and when undertaking the appropriate activities in the room.

Glare problems caused by lamps can usually be resolved by the use of various glare shield techniques or by installing diffusing covers. Glare shield techniques attempt to remove the lamp from the occupants' direct line of sight. Some examples of how this can be achieved are shown in Figure 5.10.

A variety of diffusing covers can be purchased to cover individual lamps. However, they should not reduce illuminance below recommended levels.

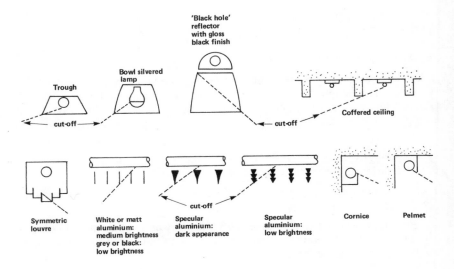

Figure 5.10   Use of Reflectors and Louvres to Control Direct
Glare (From the IES Code 1977)

Another approach based on the concept of diffusing covers is luminous ceilings. These usually consist of fluorescent tubes and incandescent lamps mounted on the ceiling with a translucent barrier between the fittings and the work area. The inside of the barrier is enclosed and therefore does not catch very much illuminance-reducing dirt. The barrier should be easily removable to facilitate lamp and tube maintenance. This type of ceiling could be useful in an installation where VDUs are to be used, but care must still be taken to avoid reflections on the VDU screen face.

Glare conditions may be created by the poor positioning of strip lighting. In order to minimise this possibility, strip lights should be installed parallel to both the viewing direction of the room occupants and to the windows. From a physiological point of view, these two requirements are quite correct, but implementing them may, in practice, result in too uniform a workroom lay-out. So, provided that the arrangement of the lamps minimises the risk of glare, these recommendations need not be rigidly adhered to. This would apply, for example, to lighting installations in which the lamps cannot be seen directly when viewed from the side, eg when the lamps are recessed into the ceiling with deep grid-type installations etc.

Windows can be a high glare source, and should therefore be fitted with curtains or blinds. Curtains should have a similar reflectance to that of the walls. All curtains should be easy to position and all blinds should be easy to adjust and clean. Glare from windows can also be controlled by the use of smoked glass but care must be taken not to give the room an oppressive atmosphere. Smoked glass windows may also be useful in reducing radiant heat from direct sunlight which becomes unbearably hot for the room occupant it falls on.

F Assessment of Reflection Levels

Reflection levels are quantified in terms of the reflective capacity of a surface and are measured in terms of reflection factors. Reflection factors indicate the proportion of the amount of light falling on a surface that is reflected and so can have possible values between 0.0 and 1.0. Reflectances of walls and surfaces play a large part in determining whether a room seems pleasant or not. If the recommendations for reflectances and illuminances in Figure 5.11 are adhered to, a room should be pleasing to a majority of people.

Generally speaking, lighter colours are preferable for walls. Saturated colours should be avoided on large surfaces as they can be distracting. Walls should be matt in appearance to avoid reflections. If walls are illuminated by daylight during the day, the artificial lighting provided for the evening and night should match the daylight effect.

Ceilings should have a high level of reflectance; white ceilings serve this purpose.

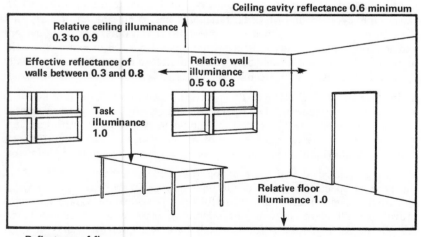

Figure 5.11 Recommended Ranges of Reflectances and Relative Illuminances for Room Surfaces (From the IES Code 1977)

The reflectance of the floor cavity should be between 0.2 and 0.3. The floor cavity consists of any surface below the working plane. Strong saturated colours should be avoided. The light reflected from coloured floor surfaces can ruin an otherwise well-balanced lighting environment.

It is advisable, though not always possible, to use furnishings and equipment of low reflectance. For desk tops, matt finishes with a reflection factor of 0.4 (with 0.6 as a maximum) are recommended.

G Review of New Lighting Scheme

If any changes have been made as a result of assessing the existing lighting, the occupants should be asked their opinion of the new lighting scheme after it has been installed and working for 1 or 2 months.

Lighting — Eyesight

In recommending lighting schemes for interiors, it is presumed that the operatives' eyesight is good or has been corrected. A survey conducted by the Association of Optical Practitioners has shown that about one third of the UK employee population had uncorrected or insufficiently corrected visual defects.

The increasing use of VDUs in offices has required VDU operators to perform more exacting visual tasks than they have been hitherto used to and this has highlighted any existing visual defects. This has led to rumours that VDUs cause poor eyesight and/or visual fatigue. It has to be said very clearly that there is no objective evidence to support these rumours. The only evident explanation is that already existing visual defects have been exposed, by the nature of VDU work. It would be advisable therefore for employees, especially those who will use a VDU, to have their eyesight checked at regular intervals.

In 1978 the VET (VDU Eye Test) Advisory Group was formed to establish a set of tests suitable for VDU operators. The record sheets to be completed by the subject and by the tester are shown in Appendix 7. Although it is advisable to have the tests conducted by an optician, the majority of the tests can be undertaken on a commercial vision screener by trained lay personnel.

Having established that some people may need spectacles for VDU work, the type of spectacles required will obviously depend on the type of visual defect. Spectacles that might be useful for VDU work are varafocals which provide for all ranges of vision between long distance and reading distance. An alternative could be trifocals which have segments for near vision, far vision and an intermediate distance.

The decision as to who should pay for the spectacles for VDU work is being debated at the moment. Management and unions should discuss these issues to establish a policy on eye tests and spectacles.

The dominant eye was mentioned earlier in this section: one of the two eyes is used to provide the main view of the visual field, this is the dominant eye. The other eye is used to provide additional detail which gives depth perception. The dominant eye also dictates location of the eyes, focal length, etc, the other eye following as a 'slave'. If a person's workstation forces him/ her to use the 'slave eye' in a dominant way, this can produce discomfort and affect performance. It is therefore important to provide flexibility in the workstation so personnel can use their dominant eye accordingly.

ROOM CLIMATE

Factors Affecting Room Climate

The subjective feeling of comfort or discomfort of room occupants is determined by the following environmental factors:

- air temperature;
- air humidity;
- air flow and speed;
- radiant temperature;
- level of cleanliness.

The first four factors interact to produce an overall sensation of comfortably warm, cool, hot etc. They cannot be considered in isolation, since they each affect the other. For example, in offices with steel walls, in winter, where these walls have a low temperature, the required air temperature will be considerably higher, than for a brick and plaster walled office. This is because the warm bodies of the occupants are losing considerable quantities of heat to the cold metal walls by radiation. This radiant heat loss has to be off-set by a corresponding increase in the convective heat gain, hence requiring a higher air temperature. The inverse is also true.

Air Temperature

Air temperature is one of the key determinants of personnel comfort. The temperature for comfort will depend on the level of physical work being expended. The greater the physical exertion in a task, the more metabolic heat is generated within the employee's body. This means a lower air temperature is regarded as comfortable, than in the case of the sedentary worker, where little energy is expended and little metabolic heat generated. Also people vary considerably in what they regard as comfortable. It is never possible to please all of the people all of the time, but some guidelines are available. For light office work, the Offices, Shops and Railway Premises Act specifies a minimum temperature of 16° Centigrade, below which work cannot be expected to be undertaken (legally). However, a more reasonable range for comfort is 21° − 23° Centigrade.

Humidity

Relative humidity is a measure of the amount of moisture in the air at the existing temperature, and is expressed in percentage terms. The normally accepted humidity comfort range is between 30 and 70%. However, for office environments, the relative humidity should be kept above 50% and should not fluctuate widely. The symptoms of a very low humidity level, particularly if combined with a high temperature, are dry noses and throats. The symptom of too high a humidity can be stuffiness.

Air Flow

At the above recommended temperature levels the rate of air flow should be about 15cm/second. This rate is just about perceptible. Lower levels of flow may be perceived as 'airless' and higher levels as 'draughty'. The cooler the air, the more its movement can be detected. One way of removing the discomforting aspects of draughts is to raise the air temperature, making the draught less perceptible. If the air temperature is too high for comfort, but cannot be reduced, the overall feeling of comfort can be improved by increasing air flow. This increases evaporative heat loss through sweat evaporation, and makes the person feel cooler, and thus more comfortable. This can be particularly helpful in very hot summer conditions, when the use of fans is recommended.

Radiant Temperature

Radiant temperature is a measure of the heat emitted by a surface or body. When a person works in a room with equipment, he or she will be radiating heat to everything else − walls, floor, ceiling, table, equipment, etc, and receiving radiant heat from all these items. Where the radiant temperatures are close to the desirable air temperature − say 20°C − they will not greatly influence comfort. If, however, they differ considerably (± 50%), they will affect comfort quite considerably. This can be offset, as previously mentioned, by appropriate changes to the air temperature. Getting a reasonable balance is best done by experimenting and seeking the subjective response of those working in the area.

Cleanliness

The comfort of a particular atmosphere is affected by the presence of dust, fumes or smells in the air. Some dust and fumes can not only be irritating but even extremely dangerous. In general terms, offices are far less likely to encounter such problems than shop floors in factories. However, attention should be drawn to the following:

- offices located within a factory site may receive industrial dust, fumes or smells through open doors and windows, or through air conditioning systems;
- offices located close to roads, loading bays or vehicle compounds and garages may be subjected to exhaust fumes via open doors and windows or through air conditioning systems;
- certain types of chemicals used in office equipment such as duplicating machines may give off harmful fumes.

It has also been suggested that the presence of positive ions in the air we breathe causes irritability and tiredness.

The Prevailing Room Climate

The most common causes for complaint about room climates are:

- being too hot;
- being too cold;
- draughts;
- stuffiness;
- direct sunlight.

These problems are often caused by one or more of the following physical characteristics of a room:

- the heating system;
- doors;
- windows;
- radiant temperatures (a more uncommon cause).

Some basic information about each of these room characteristics should have been established during the collection of environmental data.

In order to gain a broad understanding of the prevailing room climate the following points should be investigated:

The heating system

- how the system works;
- under what conditions it comes into operation;
- who is responsible for the system;
- how it can be controlled.

Doors

— are they kept open or shut?

— when open, do they cause draughts or let in air of different temperature, cleanliness or humidity?

Windows

— are they easy to open or shut?

— if they have blinds are they easy to adjust?

— are they kept open?

— when open do they cause draughts or let in air of different temperature, cleanliness or humidity?

Radiant temperature

— are there any particularly cold surfaces or objects?

— are there any particularly hot surfaces or objects?

— if yes to either of the above questions, determine duration of cold/hot, frequency of occurrence, surface area concerned and its distance from personnel, and radiant temperature.

Having gained a broad understanding of these four physical characteristics of the room, the subjective assessment of the occupants should be obtained. In particular the occupants should be asked:

— if they are sometimes too hot?

— if they are sometimes too cold?

— if there are draughts?

— do they find the room stuffy?

— are they ever troubled by direct sunlight which cannot be blocked out?

Changing the Room Climate

Computer terminal equipment can affect room climates though the effect is likely to be limited to temperature and air flow. Information about how much heat the equipment emits and what level of air flow its cooling system is likely to generate should be provided by the manufacturer. If a full-scale study of the room climate is being undertaken by a heating engineer, these figures can be included in the calculations. If this is not the case, it would be advisable to observe the terminal in operation in another installation and to talk to the users about any heat- or draught-creating effects.

Room climates are extremely difficult to regulate satisfactorily. It is usually impossible to completely satisfy all those working in the space concerned. Inevitably some complaints will be received no matter how sophisticated the room climate control systems are. It is possible to spend very large amounts of money on sophisticated heating and environmental control systems without gaining significant improvements. Therefore a great deal of discretion is required in reaching a conclusion about the necessary changes.

For this reason it is not advisable to undertake anything but minor changes (such as putting draught excluders on doors or windows, or installing window blinds) without consulting a heating engineer. He will be able to consider the need to change the heating system, the installation of air conditioning complete with filtration system and humidifier, or even the application of IED — integrated environmental design — which takes into account the amount of heat generated by people and equipment.

Whatever situation applies, the following basic human aspects should always be remembered:

— where provision is made for localised control of the heating system, the room occupants should be trained in how to operate the controls;

— if blinds are provided they should be easy to adjust without having to take up an awkward position. They should be effective in keeping out strong sunlight. They should be cleaned regularly;

— some windows should always be present — no matter how small — as they provide people with an important link with the outside world;

— in environments where the windows are expected to be opened and closed, this should be easy to do without having to adopt awkward positions and without the window itself being a safety hazard in the process.

No matter what measures have been taken to control the room climate, it is the perceived status of that environment by the occupants that counts. Therefore, after the terminal equipment has been installed, the subjective assessments of the occupants should be obtained. It should be remembered that a thermal environment must be checked over a twelve-month period to take into account variations in the seasons.

NOISE

Introduction

A noise is a fluctuation in air pressure. The rate (or rates) at which the pressure fluctuates gives the noise its particular frequency (or frequencies). The fluctuation is measured in cycles per second (c/s) and the size of the fluctuation indicates the amount of energy involved, and this is measured in decibels (dB). All noise consists of a variety of sound pressure fluctuations at different frequencies. Therefore in order to measure noise accurately it is usually recorded and later analysed to determine the sound pressure level in a number of frequency bands. The human ear does not 'hear' sounds as equally loud for the same sound pressure level at different frequencies. Some frequencies seem 'louder' than others. A special sound-measurement scale which takes into account this human perception of sound uses the unit dB(A), denoting the 'A' weighting, as it is called. This measure is used in this book.

Noise above certain levels and of particular types can cause lack of concentration, irritability, stress, and even loss of hearing. A British Government Code of Practice specifies an upper limit of 90 dB(A) above which some form of ear protection must be worn. Office work requires a much quieter environment and it is generally agreed that a noise level of about 65 decibels is acceptable in such an environment for non-auditory activities. However, where the activity has a large auditory component (eg using the telephone) 55 dB should not be exceeded.

Noise in the office is caused by:

— people talking on the telephone;

— people talking across the office to each other;

— loud, and long, telephone rings;

— doors banging;

— a regular flow of people along a hard floor;

— air conditioning equipment;

— typewriters.

Noise levels in offices may vary from hour to hour, day to day or even month to month. Consequently a true picture of an office's noise conditions will not be gained by making an assessment over a period of a few hours. Therefore, in addition to making such an assessment, the room occupants should be asked their views on the room noise conditions, and asked to identify any particularly irritating noise sources.

Different types of computer equipment generate varying amounts of noise. Generally speaking, VDUs are fairly quiet whilst hard copy printers are very noisy. Working noise levels of particular pieces of equipment are usually specified by the manufacturer. However, such figures can only be used as a guide to the final noise generated by the equipment when it is in use. This is because:

— the manufacturers' figures may only reflect average rather than peak levels;

— the equipment mountings may affect the amount of noise generated;

— different rooms have different levels of noise absorbency;

— while the equipment itself may have a relatively low noise output, it may combine with other sources of noise in a room to create an undesirable noise condition.

The noise in any office is generated by primary sources, already described, and by secondary sources, which reflect sound incident on them. The main secondary sources are walls, floors, ceilings, table tops and metal office equipment, such as metal cupboards (6' x 3' for example). The shape of a room can also affect the noise levels. For example, a long narrow room with rows of noisy equipment will have a higher noise level than a square office with the same floor space and equipment. This is because the sound from primary sources has a short distance to travel to the wall to be reflected back into the room, and so is at a higher level hitting the wall, and a higher level coming back into the room. Low ceilings can also bounce back a lot of reflected sound. If walls, floors and ceilings have very hard surfaces they will reflect a high proportion of sound incident on them. This produces secondary noises, which can sound like echoes, (which they are) or reverberations. All interior surfaces should be as sound-absorbent as possible.

Noise Reduction

Where a serious noise problem is identified it will be necessary to call in a noise engineer to undertake detailed sound measurements and to implement changes to reduce the noise. However, in offices, it is often possible to solve the problem without recourse to extensive sound reduction exercises. In the case of hard copy printers, for instance, silencer covers can be obtained for most of the brands on the market, although such silencers should not make the paper-handling facilities too difficult to use. Other techniques of dealing with the problem include:

— isolating the equipment from the people being affected by the noise;

— reducing the noise emitted from primary sources by insulating them, housing them in booths, putting hoods over them, reducing noise from telephone rings, etc;

— reducing the noise from secondary sources, by carpeting the hard floors, placing sound-absorbent panels/material on walls, ceilings, sound-absorbent room dividers etc.

ROOM LAYOUT

Introduction

The following factors need to be taken into account when undertaking a room layout exercise:

- work groups;
- open plan vs small offices;
- space availability;
- constraints on workstation positions;
- wiring and plug sockets.

Work Groups

The importance of work groups cannot be overemphasised. The social aspects of a closely-knit work group, and the sense of belonging which is generated, can be very important in retaining a happy and productive work force. In some cases, physically positioning a group of workers together can be counter-productive in that the amount of nonproductive social intercourse becomes excessive. Obviously compromises must be reached based on a clear knowledge of the work groups that exist, the work to be done and the type of people who are doing the work. Breaking up an existing work group can result in considerable dissatisfaction, poor productivity and even resignations. Therefore, in many situations physically dispersing a close-knit group should be avoided whenever possible — particularly when the members of that group are engaged in boring or repetitive work.

Open Plan vs Small Offices

Both open-plan offices and small offices have benefits and disadvantages. However, the decision in favour of one or the other often depends on the existing layout, the characteristics of the building and the cost of conversion — all of which vary from building to building. However, the following human aspects should also be considered:

- noise is generally greater in an open-plan office, unless acoustic tiles, carpets and partitions are used to absorb it;
- glare and screen reflections from the lighting and windows are generally greater in open-plan offices and more difficult to resolve;
- some degree of privacy is required to undertake work demanding concentration;
- some means should be found whereby people's territorial instincts can be satisfied and the employee can conceptually define his own space.

Space Availability

Space is a vital factor in obtaining a successful room layout. Too much space is effectively a waste of resources; too little space will reduce efficiency, cause dissatisfaction and possibly endanger the employee's health and safety. Put in these terms the concept of space availability is easy to understand. However it is much more difficult to actually measure space as illustrated by the stance taken on space availability in the British Offices, Shops and Railway Premises Act. This act sets minimum legal requirements as shown in Figure 5.12.

Ceiling height in inches (centimetres)   90 (229) 96 (244) 102 (259) 108 (274) 114 (290) 120+ (305+)
Square feet (square metres) per person habitually employed in room 54 (5.0) 50 (4.6) 47 (4.4) 45 (4.2) 42 (3.9) 40 (3.7)
No ceilings to be below 7' 6" (229 cm) in height In rooms with ceilings lower than 10' (305 cm) in height a minimum of 400 cubic (11.3 cu metres) per person habitually employed in the room, is required. All measurements should be taken without regard to furniture, fittings, etc.

Figure 5.12 Legal Minimum British Office Space Requirements

Although it is important that such minimum legal requirements exist, in practice they are irrelevant for two reasons:

— the act requires all measurements to be made without regard to furniture, fittings etc. Therefore in practice an office could be well within the legal limit while being packed out with furniture and equipment;

— commonly accepted standards of office accommodation in 1980 are far higher than those indicated by the minimum legal requirements. Furthermore, all the signs point to a continued improvement in these standards.

Therefore if the question of space availability is to be considered objectively, the measurement procedure must reflect the true space conditions of the office, and the space standards that are set must be appropriate for the particular time and place.

There are many ways of defining what measurements should be taken. However, perhaps the most important thing is to have an agreed measurement procedure which can be used throughout an organisation so that the space availability figures calculated for different offices can be compared. An example of one possible method of space availability measurement is given in Appendix 8.

Once an agreed method of measurement is being used within an organisation, the space availability of different offices can be compared. As a result of making such comparisons it may or may not be felt appropriate to set standards. If standards are set, it would probably be unwise to adhere to them too rigidly since it is almost impossible to undertake the measuring to a high level of accuracy. Standards will probably be of more use if they are employed to:

- identify and rectify particularly poor office conditions;

- provide a guide for when offices are being reorganised or set up.

Constraints on Workstation Positions

The positioning of workstations in a room may be constrained by the following factors:

- for right-handed people, the light should come from the left in order to avoid shadows cast by the hand. Light arriving obliquely from the rear can be even better sometimes. For left-handed people the opposite applies;

- positions where the user is subjected to glare from light arriving directly from the light source should be avoided;

- workstations housing VDUs must be carefully positioned to avoid reflections of the windows or lights on the screen surface. In effect this means that VDUs should be placed at 90-degree angles to the windows and moved to a position where glare and reflection problems from the light sources are minimised. If this proves difficult, other options are to screen off the VDU workstation or to try out an antiglare filter;

- workstations should not be positioned in very hot locations (eg next to a radiator) nor where draughts are likely (eg next to a well-used doorway);

- that part of a piece of equipment which gives off heat should not be placed close to anybody. Several pieces of equipment which give off heat may cause a localised hot, dry area if placed together;

- positions next to high noise sources are unsuitable;

- when laying out a room, provision must be made for passageways of adequate width. A passageway 86 cm wide (minimum 72 cm) permits one person to pass another who has to turn sideways. For two people passing each other 130 cm (minimum 106 cm) is required. The minimum passage for one person is about 53 cm. Passageways must, of course, be kept free from obstructions at all times;

- space must be available to allow doors to be fully opened and shut;

— space must also be available to allow windows to be opened or shut without difficulty; for blinds to be adjusted and for windows and blinds to be cleaned;

— the space to undertake equipment maintenance must be easily obtainable if required;

— easy access to fire fighting appliances must be available at all times.

Wiring and Plug Sockets

A large amount of electrical equipment is used in offices, the most common examples being:

— the telephone;

— the mains-powered calculator;

— the electric typewriter;

— the visual display unit.

All such equipment is usually plugged into circuit sockets via wiring. When laying out a room, the position of the sockets must be identified and then provision made for the routeing of the wiring. Both activities are subject to health and safety constraints.

Trailing cables along or across a passageway are safety hazards. To try and eliminate this hazard people often consider placing the cables under a carpet or under some kind of cover fixed to the floor. The former solution is unsatisfactory because it affects the fit of the carpet and is a possible fire hazard, and both approaches still represent a trip hazard. One other apparent solution is to run wires across the ceiling and dangle them to the appliance. This certainly does represent a safety hazard and should not be attempted. A much more satisfactory solution is to lay out the room so that the cables will always be alongside or behind the furniture or next to walls. In practice, this means either locating desks near to wall-mounted sockets or placing sockets in the locality of groups of desks.

The health and safety difficulties described above, and the prohibitive expense and disruption of relocating or installing more sockets, often greatly reduce the amount of flexibility available in the office layout. This is undesirable since organisations and work requirements are changing increasingly rapidly, and demand an equivalent degree of office layout flexibility.

The best solution available — to place socket points, sunk in the floor, at very close intervals throughout the floor area — is very expensive. The cost can be reduced by increasing the floor area served by each socket unit; and flexibility can be maintained to a certain extent by provisioning for the installation of more sockets in the future. However, even this is not a cheap option.

Designing the Layout

There is no one foolproof method of designing a workstation layout. However, two methods are mentioned here, perhaps at opposite ends of the scale in terms of complexity.

Firstly, the method of link analysis can be used. The word 'link' refers to any connection between a person and a piece of equipment, or between one person and another. Links include walking, talking, seeing and movement of information. The importance of each link, and its frequency of use must be established. The relative values of each link are formed from some suitable composite of these two values.

A link diagram is then drawn with circles representing people (each job having a code number), squares representing equipment or parts of them (each uniquely identified by a code letter), and the lines between any of these representing the links. The links are drawn with the higher values having the shorter links, and so that there is a minimum of crossing links. This is the optimum link diagram. A scale drawing of the office should then be made with the equipment in the locations suggested by the link diagram. If the office concerned is very large with many items of equipment and personnel, link diagram computer programs exist which carry out this exercise rapidly

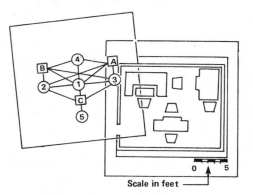
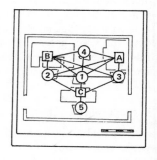

Scale in feet

The circles represent people in the system, with a code number for each job. The squares represent items of equipment each uniquely identified by a code letter

Figure 5.13 A Link Diagram Superimposed on a Scale Drawing
(Taken from Van Cott and Kinkade)

and efficiently.

Secondly, a simpler method is to make a scale drawing of the room and shade all the constrained areas. Scale cardboard cutouts of the items to be fitted in the room can then be placed on the scale drawing until a suitable possibility is found. Magnetic strips on a magnetic board are an alternative to cardboard cutouts.

When a possible alternative is identified it should be evaluated against:

— any work group requirements that were established earlier in the room layout process;

— any constraints on workstation position that were established earlier in the room layout process;

— health and safety criteria for the routeing of wiring.

When two or three possible alternatives have been identified, the potential occupants should be asked their opinion of the options and their comments taken into account when the final decision is taken.

Once the final layout has been agreed, plans must be made to implement it. The disruption caused to the occupants must be realistically estimated and catered for.

After the workstations have been positioned, it is essential to identify each piece of new equipment. This will enable equipment maintenance and its associated documentation to be controlled. The identifier, be it alpha, numeric or alphanumeric, should be unique and meaningful to users, systems people and maintenance personnel alike. In many cases, people tend to refer to the equipment by its geographical location, and therefore this would be an appropriate identifier. Any reference to the terminal in computer output should always bear this identifier as opposed to some other internal machine identifier. Each terminal should be clearly labelled with its own identifier. When an identifier has been assigned to a particular terminal, it should be listed on a Terminal Identification Record (see Appendix 9) which should be held by an appropriate central body. There is no guarantee that the final agreed layout will be satisfactory in practice; therefore a review of the layout should be undertaken after it has been in use for 2 or 3 months. Such a review should take into account the views of the room occupants.

UPKEEP OF EQUIPMENT, WORKSTATIONS AND WORKPLACES

Introduction

The upkeep of equipment, workstations and workplaces is, in effect, a maintenance operation. There are two types of maintenance: planned maintenance and breakdown maintenance. Because other arrangements can be made for the period when planned maintenance is being carried out, and minor faults can be prevented from escalating into serious major faults, in most circumstances it is better to spend money and effort on planned maintenance in order to avoid breakdown maintenance. For both types of maintenance, a record should be kept of the work undertaken; this will assist fault-finding and the prediction of problems, and can act as an input to the purchasing of replacement equipment.

Advice about the technical engineering maintenance of terminals is often provided by the manufacturer. However, the upkeep of those aspects of the machine, workstation and workplace which ensure efficient use of the equipment by the user is equally important. In effect, this means the regular maintenance of:

— the equipment;

— the workstation;

— the lighting;

— the room climate;

— noise in the room;

— the room layout.

Maintenance Checks

In order to carry out meaningful maintenance on these items, specific checks or specific work must be defined so that a required standard is maintained or the need for further maintenance work is identified.

The checks that are defined will vary from installation to installation. However, a typical list for an ordinary office VDU/keyboard installation is given in Appendix 1.

The Responsibility for Maintenance

If the maintenance of equipment, workstations and workplaces is to be done effectively, the responsibility for this maintenance activity must be allocated to a specific job title.

It should be the responsibility of this job title to:

— hold the installation instructions, a technical specification of the machine, and maintenance instructions;

— hold and update the Terminal Identification Record;

— identify the checks that must be made, or work that must be undertaken, to carry out the planned maintenance activity. Also specify the time period between planned maintenance inspections for each check. For technical engineering maintenance the manufacturer will provide this information. For upkeep maintenance, the checks must be identified according to the requirements of each installation, and the most appropriate time periods between planned inspections established by experience;

— prepare planned maintenance schedules;

— make arrangements for planned and breakdown maintenance to be carried out, and ensuring that it is carried out satisfactorily;

— ensure that details of all planned and breakdown maintenance carried out are recorded;

— monitor the maintenance records to identify any inherent weaknesses or likely future breakdowns.

The User's Role

When undertaking terminal, workstation and workplace maintenance activities, it is essential to solicit the view of the equipment operators and room occupants as well as making the relevant inspections. Some full-time users of terminals might welcome some responsibility for care of their equipment, workstations and workplaces. Users who are given such responsibility must be quite clear about:

— what aspects they are responsible for;

— how they record details of the maintenance checks they make;

— what action they can take;

— what information they must feed back to the person in overall charge of maintenance.

Where users are not given responsibility for upkeep maintenance, it is essential to solicit their views during the relevant maintenance inspections.

6   User Support

PLANNING USER SUPPORT

Introduction

A user support network provides computer system users with sufficient knowledge, skill and confidence to enable them to use the system efficiently. It is unlikely that any one approach to user support will be sufficient, and therefore the most appropriate combination of methods must be implemented for each particular computer development.

The main methods of providing user support are:

— training;

— documentation (ie manuals, etc);

— human support (ie people who will help users);

— change mechanisms (ie methods by which a user can suggest ways in which the system could be improved).

The purpose of planning these user support facilities as a network is to ensure that each mechanism complements the other, and that the network as a whole meets all user support needs. In order to conduct this planning exercise, a certain amount of initial design work for each mechanism must be undertaken.

Planning Stages

There are five stages in planning a user support network:

— Establishing the data required to plan a support network;

— Planning human and documentary support;

— Planning change mechanisms;

— Planning a training programme;

— Assessing the support network.

Establishing the Data Required to Plan a Support Network

The following will be required:

— a list of potential users of the system;

— for each person on the list of potential users the following information should be established:

— details of any formal qualifications;

— job experience;

— job experience of computers and computer systems;

— experience of computer terminal equipment.

This information will ensure that any training programme will not cover material already familiar to the user, but will build on his capabilities.

The following will also be required:

— details of the activities that the users will have to carry out. This inform-ation will help the training programme designer to establish in what subjects the users need training;

— details of any existing human support mechanisms (computer advisory personnel, local experts, managers/supervisors, human interfaces, organ-isational representatives). This information will be required to decide whether existing human support mechanisms are satisfactory;

— details of any existing documentary support (manuals, circulars, 'help' facilities in existing systems, computer-based documentation) and of any documentation standards in use within the organisation. This information will help the designer to decide on methods of document-ary support;

— details of any existing change mechanisms (suggestion schemes; person-al conversations with, or letters to, supervisors/managers; departmental meetings; input to systems monitoring committees; computerised change systems). This information will help in deciding whether exist-ing change mechanisms are satisfactory or if new mechanisms will be needed.

Planning Human and Documentary Support

Human and documentary support complement each other and should there-fore be planned together. Generally speaking, both human and documentary support meet short-term user needs for 'help' information. Therefore they are the support mechanisms which users will encounter most, and around which other support mechanisms should be built.

Planning Change Mechanisms

This may be affected by the human and documentary support that will be available (eg if supervisors are to be the prime source of human support it may also be appropriate to have users send their ideas for system changes to them; or if computer-based documentation is to be used, it may be appropriate to have a computerised change system). Consequently change mechanisms should not be planned until the human and documentary support measures have been specified.

Planning a Training Programme

The training programme should not be planned until the other user support measures have been planned because the training programme will either teach about, or actually use, the other user support material.

Assessing the Support Network

A clear idea of the support network should have been established by this stage. The network should be assessed to ensure that its various parts are complementary; and to ensure that it will provide the users with sufficient skill, knowledge and confidence to enable them to operate the system efficiently. Any shortcomings should be eliminated by making appropriate modifications to the network.

TRAINING

Introduction

A training programme is a predetermined programme of work which has specified teaching objectives. The potential users of a new computer system should be given training with the overall objective of ensuring that they can use the new system effectively.

There are eight stages in carrying out a training programme:

— identify people to be trained;

— establish the needs of each individual;

— determine teaching objectives to meet the needs;

— identify the training modules;

— prepare the content of each module;

— make arrangements to carry out the programme;

- carry out the training programme;

- review of the programme.

Identify People to be Trained

The people who will be affected by the new computer system will not only be the immediate obvious users (eg stock control clerks in a stock control system) but also the less frequent or physically removed users (eg auditors or the general public in a billing system). The existing skills and capabilities of these people must be established so that the appropriate kinds of training programmes can be devised.

Establish the Needs of Each Individual

Training needs can be broken down into two broad categories:

- basic introductory needs;

- skills and system training needs.

Basic introductory needs are usually common to all potential users and include the need to:

- gain a broad understanding of the purpose and major functions of the system, and its interfaces with other systems;

- become acquainted with the terminal equipment, control equipment and connecting mechanisms which make up the computer hardware. This should include seeing the equipment and getting a broad idea of how it works, eg by taking the potential users to see the equipment on the manufacturer's premises, before it is purchased. This approach may also produce a much more positive user attitude towards the new system;

- learn how changes to the system can be suggested or instigated.

Skills and system training needs depend on individual requirements and must take into account existing skills and capabilities. The training programme should cater for older, more experienced staff who will probably learn more slowly and have a less positive attitude towards change than younger staff. In general terms most users will want to learn:

- what the system can do for them;

- how to use the system;

- how to use the equipment;

- how to integrate the computer activities into their existing work roles;

- how to extricate themselves when things go wrong;

— how to use the various facilities provided in the user network.

The training needs of a hypothetical stock control clerk are given in Appendix 10.

Determine Teaching Objectives to Meet the Needs

Each identified need should be considered separately and a teaching objective set to meet it. Each objective should be capable of being met within the time available and given the level of ability of the trainee. Whenever possible each objective should be measurable to provide the trainee with assessments of his progress; knowledge of results is extremely important in the human learning process. Some examples of teaching objectives for specific needs are shown in Figure 6.1.

NEED:	To know what information is held for each stocked item (eg stock on order, stock on hand, etc): and to understand how each of the pieces of information is obtained (eg stock on order is updated when a production order for that item is input to the production system)
OBJECTIVE:	To be able to work out the value of each type of record given a predefined set of circumstances
NEED:	To know what each key on the keyboard does
OBJECTIVE:	To be able to press the correct key(s) in response to instructions given on the screen

Figure 6.1 Example of Teaching Objectives for Specific Needs

Identify the Training Modules

The training programme should be made up of a number of self-contained modules. To establish what these modules are, the various objectives set for each individual must now be inspected to identify duplicates and overlaps. Similar objectives should be grouped together if they are suitable for being covered in the same module. Some modules may cover a large number of objectives common to most trainees. Others may be specially designed to meet the needs of individuals. Some modules may require several separate sessions — others may require only one session of relatively short duration. Some typical modules are shown in Figure 6.2.

Prepare the Content of Each Module

Once the modules have been determined, the general techniques and media to be employed must be established. Possible techniques include:

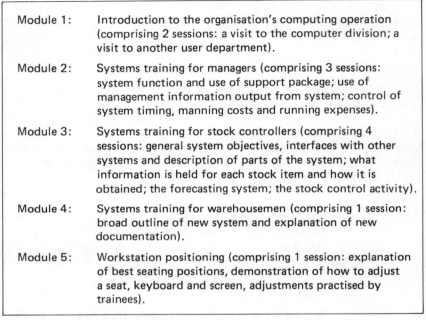

Module 1:   Introduction to the organisation's computing operation (comprising 2 sessions: a visit to the computer division; a visit to another user department).

Module 2:   Systems training for managers (comprising 3 sessions: system function and use of support package; use of management information output from system; control of system timing, manning costs and running expenses).

Module 3:   Systems training for stock controllers (comprising 4 sessions: general system objectives, interfaces with other systems and description of parts of the system; what information is held for each stock item and how it is obtained; the forecasting system; the stock control activity).

Module 4:   Systems training for warehousemen (comprising 1 session: broad outline of new system and explanation of new documentation).

Module 5:   Workstation positioning (comprising 1 session: explanation of best seating positions, demonstration of how to adjust a seat, keyboard and screen, adjustments practised by trainees).

Figure 6.2 Some Hypothetical Examples of Training Modules

— classroom-type sessions;

— lectures;

— programmed instruction text;

— simulations;

— audio visual presentations;

— taped lectures;

— computer-assisted learning;

— visits;

— practicals;

— demonstrations.

The techniques chosen will depend not so much on what is possible — since any number of techniques, or variations and combinations are possible — but more on the resources available and the objectives to be met.

Another constraint is the need to choose a technique which will suit particular individuals. For instance a busy manager may prefer to listen to taped lectures on his way home; or for large numbers of the general public a short written communication backed up with televised advertisements may be appropriate.

When appropriate techniques have been determined, the actual content of each module must be planned and tested. Three points should be remembered when planning modules:

— the trainee should be active during the learning process, and, where possible, getting hands-on experience;

— the trainee should be able to ask questions and get responses during the learning process;

— regular feedback on his progress will help the trainee.

Making Arrangements to Carry Out the Programme

After the content of the programme has been planned, the lengths of the modules and the resources required to carry out the programme should be known. A detailed administrative effort must now be undertaken to arrange times, dates, places and people, to allow the programme to be carried out. Generally speaking, the training programme should take place just before the knowledge or skills it is imparting are to be used in practice. However, some leeway must be left to allow for the inevitable slippage that will occur in carrying out the training programme.

With the exception of clerks (engaged in full-time data entry) and some specialists, computer usage is not generally a central job interest for most users. This means that the potential users may not be as keen to learn how to use the system as they might be to learn skills or to gain qualifications directly related to their job function. It also means that gaining sufficient time for adequate training may be difficult to justify. These problems must be dealt with before the training programme gets underway.

Carry Out the Training Programme

Carrying out a training programme is like any other planned activity; it needs monitoring and controlling. Things are bound to go wrong. Sessions may have to be rescheduled to cater for unforeseen demands on the trainees' time. The content of some sessions may have to be changed to cater for late changes to the systems design. Some trainees may not be able to attend certain sessions, and others may not meet their objectives. All these problems and many others must be dealt with in such a way that programme objectives are achieved as far as possible within the constraints of time and other resources.

Review of the Programme

The experience gained in devising and carrying out a training programme is very valuable. It can be fed into similar programmes to ensure that the same mistakes are not made twice. Therefore when a training programme has been completed, it should be reviewed. The review should consider those aspects which were less than totally successful and record the reason why. Courses of action which can be seen to have been more successful in retrospect should be noted.

The review also identifies any outstanding or future training requirements for the trainees. Such training may be required before the system is installed, in which case a new training programme must be devised. Longer term training requirements after the system has been installed are discussed in the section on Continuing User Support (p 134).

DOCUMENTATION

Introduction

Documentary support at its best provides quick and positive assistance to help the user carry out his job. At its worst, documentation slows people down, thwarts their attempts to learn, confuses and frustrates them. Five documents are needed for any piece of computer terminal equipment:

- installation instructions;
- maintenance instructions;
- technical specification of the equipment;
- users' guide to operating the equipment;
- users' guide to operating the system.

The first three documents should be supplied by the manufacturer. They will be of little interest to the user once the terminal has been installed and should be held together for access by the person responsible for equipment upkeep.

Users' Guides

The users' guide to operating *the equipment* should also be supplied by the manufacturer. This provides instruction about use of the physical controls of the machine and should always be in full view on the workstation. Such a document should not be regarded as definitive. It may require modification to explain things in more detail or in clearer language. Other information such as the location of the user guides can also be included if required.

The users' guide to operating *the system* is usually produced by the systems designers. However, two valuable benefits in having the user actively involved in, if not solely responsible for, the production of this guide are:

— a member of the user group is forced to become thoroughly acquainted with the system;

— the guide is more likely to be produced with the terminology, style and emphasis which is understandable to the user population.

The following points should be borne in mind when producing user guides:

— the information should be presented in a logical manner so that it is easy to access;

— the information should be concise. Use of the imperative — 'do this, do that' — will help to ensure conciseness;

— the layout should be such that the information required can be identified easily and read clearly;

— any documentation standards that exist in the organisation should be adhered to.

There are two main types of documentary support:

— paper-based documentation;

— within-system aids.

Paper-based documentation can be provided either in the form of manuals or circulars. Manuals are often cumbersome and bulky which makes them difficult to use. In general, users only require part of the total set of facilities provided by the computer, and therefore it can be advantageous to provide an instruction booklet which is custom-designed for particular user groups rather than to issue the same manual to all users. However, such booklets cease to meet user needs after an initiation phase has passed if their content concentrates on operation of the system to the exclusion of general information about the system. Most manuals cannot be used without prior instruction or explanation since users need to know how to formulate their queries, how to use the index, and how to interpret the meaning of the contents. As a result, most manuals cannot be used by totally naive users and should not be regarded as self-teaching. If it is important to provide for self-instruction, extra investment in the manual design will be required.

Manuals will usually need to be updated during their lifetime and it is important that the physical presentation of the documentation permits easy update. Manuals are very susceptible to becoming out-of-date due to a combination of two different problems. Firstly, documentation changes are often the last thing to be completed when a systems change is made. Therefore new pages for a manual may not arrive for several days, weeks or even months after the change has been made. Secondly, after the new pages arrive at the appropriate user location it may be some time before they are inserted in the manual, and even then they may be inserted in the wrong place. To overcome these problems a sense of conscientiousness must be instilled in those people responsible for writing new documentation and those users responsible for physically updating the manual. The allocation of responsibilities for documentation maintenance to one job title may also help to alleviate these problems. In addition, the job holder concerned could also ensure that regular checks are made for lost, torn, dirty or otherwise degraded documentation.

Circulars used to notify users of changes to the system often remain unread, are thrown away or are lost. They are poor support because they are standardised and general, and are not specific to the different types of system user. Users should only be sent a circular if its contents are likely to be understood by them and are relevant. The contents of any circular should clarify which users should attend to the communication and how their work will be affected. The one-way nature of the communication denies users the opportunity to query the implications for their particular jobs or to confirm their interpretation of the contents. The user may often find it necessary to seek clarification from colleagues or systems personnel, ie from human support mechanisms.

Within-system aids allow the user to obtain information about a system direct from the computer. Some aids allow the user to ask the computer what it is doing, how it is doing it and why. Systems which have such comprehensive facilities are known as *expert systems*. By 1980 expert systems were only at an early stage of development. However, they have great potential and could become a widespread and powerful tool by the end of the decade.

Similar, but less comprehensive, within-systems aids are 'help' facilities where the user keys in 'HELP' or presses a special 'help' key when he does not understand part of the dialogue or what to do next. Typical examples of help facilities are guidance on how to respond to error messages, and the provision of prompt lists which give meanings of codes and abbreviations. Ideally, help facilities should be available to the user at any time during his dialogue with the computer. In practice it may be too difficult and costly to cater for all eventualities. It may also be feasible to ask users to identify new help facility requirements throughout the life of the system. The section on dialogue design should be referred to if help facilities are to be designed.

Another type of within-system aid is computer-based documentation. This is the straightforward replacement of paper-based documentation by an on-line documentation system. The great advantage of such an approach is the control gained over the updating exercise and the ease by which it can be accomplished. The use of computer-based documentation allows the person responsible for documentation maintenance to ensure that the information is absolutely up-to-date and always available to the users. However, before a computer-based documentation system is installed, care should be taken to ensure:

— that sufficient terminals are available to cater for users doing their normal work at the terminals as well as accessing the documentation;

— that it is possible to access the documentation in the course of normal screen work, in a way that will assist the normal work rather than hinder it. In this respect it may be beneficial to hold the normal work on half the screen while accessing the documentation on the other half;

— that the documentation is structured in such a way that it is easy to access.

HUMAN SUPPORT

Introduction

Human support mechanisms provide users with a human source of assistance, should they require it, in their use of a computer system and should be designed to complement documentation. Five major types of human support mechanisms exist, each one meeting a particular set of needs:

— computer advisory personnel;

— local experts;

— managers and supervisors;

— human interfaces;

— organisational representatives.

Computer Advisory Personnel

Computer advisory personnel have formal responsibility for assisting users. They are generally responsible to the computer department, but may be located elsewhere, either in a central location or within a user department. Although their advisory role is recognised and part of their work load is officially allocated to that function, there are, nevertheless, two disadvantages from the user's point of view:

— problem formulation may be difficult for the user as he has to state his query in a form which is meaningful to the advisory officer. If the officer is located within the user department on a permanent basis, this may not be a serious problem since he will become aware of the user's needs and language. Where advisory personnel are located away from the user department a very real 'language barrier' may exist. Furthermore, even though users may recognise that it is legitimate to approach advisory personnel for assistance, there may sometimes be a reluctance if queries are felt to be trivial, or if the user feels he ought to know the answer;

— the high level of expertise of the computer liaison personnel can also be a problem for users. Clerks and managers may be fearful of being 'baffled by science' and may feel that responses to their queries are too 'technical'.

Therefore if computer advisory personnel are to be used they should:

— be physically located in the user department;

— ensure that their responses to queries are in terms that the user can understand;

— ensure that their relationships with the users are such that the latter will not be afraid to ask advice no matter how trivial or straightforward the problem seems to be.

Local Experts

In most groups of users, advice and information about all aspects of their work, including operation of the computer, is being constantly exchanged. Effectively the users are supporting each other on a day-to-day basis. However, some of the questions users ask each other will require a more detailed, but not necessarily more technical, level of knowledge. Such assistance can often be quite adequately met by a *local expert* — an ordinary user who knows that bit more about the computer system. Local experts often arise spontaneously when one member of a group has the capacity and interest to expand his knowledge of the computer system. This individual soon functions in the role of expert adviser to his fellow users. A local expert is often a key figure in ensuring effective use of the system. The important characteristics of the local expert are his ready availability when a problem arises, and his familiarity with the user's requirements, language, task goals and job content. Many minor but potentially critical difficulties can be resolved rapidly by the local expert without involvement or even awareness of management or the systems staff. Instantaneous advice on trivial, but essential, actions such as how to log-on, reminders of operating procedures for the intermittent user, interpretation of error messages, translation of output, etc, can be provided daily by local experts.

If a local expert is to fulfil a human support function, the following steps should be taken:

— encourage any potential 'local expert' by giving him opportunities to learn more about the system. Offer him friendly advice or instruction about the use of the computer system. Minimise use of jargon and give clear explanations of the system at an appropriate level of complexity;

— in the absence of a local expert evolving in a group, select one member of the group who can benefit from special training about the system. Equip him with knowledge about the range of appropriate facilities, of reasons for common system faults and malfunctions, and about future plans for system development;

— formally recognise the local expert's role.

Managers and Supervisors

The users' most common source of help on work matters is the manager or supervisor, and this can quite naturally be extended to include computer work as well. However it does require the manager or supervisor to learn not only as much about the computer system as the people under them, but more. One way of achieving this is to first train managers and supervisors in the details of a new computer system and then to have them train their own people. In organisations where this is a new approach, managers/supervisors will require training in computing fundamentals and in how to run training programmes.

If managers/supervisors are to be used as a source of human support the following steps should be taken:

— ensure that they will have sufficient time to carry out these functions in addition to their normal managerial or supervisory duties;

— ensure that they are adequately trained to carry out the function;

— ensure that their relationship with the users is such that the latter will not be afraid to ask advice no matter how trivial or straightforward the problem seems to be.

Human Interfaces

A human interface acts as an intermediary between an end-user and the computer. In this way people such as senior managers or doctors, are able to request and receive specific information, without having to go through the time-consuming process of locating the information in the data bank and requesting the hardcopy, etc. The human interface usually transforms the user's requests for computer output into a computer-compatible form such as an output request form or an appropriate file enquiry. Procedural aspects of this job such as filling in forms or operating a keyboard should be recognised clearly as formal duties by management, systems personnel and job occupants alike.

It is important to recognise the processing function fulfilled by the human interface. Where a transformation of users' information needs has to be carried out, the intermediary may have to determine the most appropriate way of expressing the stated needs. In the case of a computer terminal linked to a data base there may be a number of options available.

Some of the options will be more appropriate to the user's needs than others and some may require more expertise or more effort from the human interface. Users who use a computer via a human interface may complain that they have no way of knowing whether the interface has selected the 'best' category of computer output or the easiest or the most familiar one. The human interface can be tempted to select the course of action involving the minimum effort necessary to achieve an acceptable outcome. Clearly the criteria used by the human interface to determine the type of output requested may not coincide with the indirect user's criteria for assessing relevancy and usefulness of output. In consequence, the indirect user cannot be assured of receiving the most appropriate computer service unless the human interface is well-informed and committed to the same objectives. It is important to recognise this mediating effect of a human interface and not to regard him or her primarily as a mechanical extension of the computer system. Situations where one human interface works on behalf of several users who are in direct competition with each other should be avoided, since some users would be likely to receive better service than others.

The human interface may encounter a variety of pressures particularly when operating a computer terminal. For example, systems personnel may discourage the use of on-line enquiries which will monopolise the system and hinder the activities of other users. The indirect users (the managers, professionals, etc) are not concerned with constraints such as these and may demand particular outputs which they know are available yet which the human interface is not supposed to access on-line. Since the human interface is generally a subordinate fulfilling an ancillary function, he is vulnerable to injunctions from both 'sides' and may experience considerable stress constantly trying to resolve these conflicts.

Organisational Representatives

Organisational representatives use the computer belonging to their organisation, on behalf of members of the public. Typical organisational representatives include air-line booking clerks and bank personnel.

Contact with customers is a major part of an organisational representative's work. It can be either direct contact (face-to-face) or indirect (through the medium of a telephone or a letter). In both cases the organisational representative must fulfil the following objectives:

— he reflects the 'image' of the organisation to the customer through his concern and sense of responsibility;

— he receives and records instructions and queries from the customer;

— he acknowledges receipt of instructions from the customer, and supplies feedback on action taken to the customer;

— he provides the customer with explanations, information, advice and reassurance.

The representative often has tasks to perform preparatory to using the terminal. For example, he may need to record details which identify the customer before he can make an appropriate file enquiry. Clearly it is desirable to minimise data preparation in order to reduce the workload and reduce possible transposition errors.

However, the human short-term memory and, if the contact is by telephone, the cost of using a telephone line, are limiting factors. A query of any complexity will need to be carefully recorded on a data preparation form before operation on a computer terminal can commence. In addition, some manual method of recording information received from a customer is necessary in the event of a system failure. Such data preparation tasks are often performed in the presence of a customer or during the course of telephone contact with the customer. In these circumstances, speed and accuracy are essential for good customer relations as well as for effective task performance. Speed and accuracy in data preparation will be facilitated in the following ways:

— by a realistic staff-customer ratio which reflects a trade-off between customer needs for a speedy service and the company's need to minimise staffing;

— by providing an ergonomic workstation where all job aids are easily accessible, where supplies of necessary forms are readily available, where worksurfaces are well-lit and comfortably located and where terminal operation is facilitated by appropriate seating, lighting and positioning;

— by providing well-designed standardised forms;

— by ensuring compatibility of content and layout in the forms used with that on the screen or printout;

— by gaining the commitment of the representatives to minimising delays and inaccuracies.

The organisational representative receives output from the computer either on a VDU screen or in the form of a printout. He forwards information directly to the customer in face-to-face communication, or by telephone or by sending a standardised form filled in appropriately. In some cases a copy of the printed output from the computer is given to the customer. The output, whichever form it appears in, should be designed with the relevant end-user in mind. The organisational representative, often a clerk, with his greater familiarity with the system and daily handling of printouts has different needs from the customer who may have very limited experience as a computer user. The information printed out for the customer should be in plain, un-ambiguous, non-jargon language with minimum usage of abbreviations and unfamiliar expressions. Any codes which are used should be clearly translated on the printout, with explanatory text where appropriate.

If organisational representatives are to be employed, the following steps should be taken:

— ensure that the representatives fully understand their objectives when they have contact with customers;

— use standardised forms with abbreviations, coding and formats the same as those on the screen or hard copy output;

— provide representatives with an easy-to-use checklist of essential details to be gained from the customer, eg:

— customer surname;

— customer first name(s);

— account number;

— information required;

— address or telephone number to send feedback, etc.

CHANGE MECHANISMS

Introduction

After a computer system has been installed it is unlikely to remain static for its lifetime. The system will develop as the business changes and as inefficiencies and 'bugs' are discovered. Users who are encouraged to play a part in the subsequent enhancements and maintenance activities are likely to view the computer system as a useful job aid and be pleased to work with it and develop it. In addition their suggestions for improvements are likely to be many and valuable.

There are five major mechanisms by which users can contribute to the change process:

— suggestion schemes;

— personal conversations with, or letters to, supervisors/managers;

— departmental meetings;

— input to systems monitoring committees;

— direct input to a computerised change system.

For any of these mechanisms to be successful and effective it is essential that the users be given feedback on:

— the ideas that have been put forward, the discussions that have taken place about them and their current status;

— the details and status of all planned changes to the system.

This will ensure that the user is able to follow the progress of his own ideas, and allow him to formulate more viable ideas based on a knowledge of the total current situation.

Suggestion Schemes

Suggestion schemes are operated by many organisations and have a proven record of success despite being expensive and time-consuming to run. In such schemes ideas can be submitted by anybody in the organisation about how to improve any aspect of that organisation. Cash rewards are paid according to the value of the idea to the organisation. However, generally speaking, employees whose responsibility it is to improve their own or other jobs, cannot receive payment for ideas on those subjects. In effect, suggestion schemes are most effective for general ideas which employees would normally find difficult to get someone to listen to, evaluate and take action on. Therefore a suggestions scheme is unlikely to be an effective mechanism for users to play a part in the maintenance and enhancement of their *own* system.

Personal Conversations with, or Letters to Supervisors/Managers

The effectiveness of this mechanism depends a great deal on the supervisor/ manager concerned. He must be approachable and must encourage users to come forward with ideas. The supervisor/manager should then take the matter further and actually initiate the changes.

Departmental Meetings

Regular departmental meetings are often used as a forum for discussing the computer systems operated by that department. For this mechanism to operate effectively the following conditions should exist:

— users must be encouraged to come forward with their ideas and must be confident that their ideas will not be ridiculed or dismissed without serious examination;

- a decision must be made about each idea put forward, ie it must be either accepted, accepted for further investigation, held over until the next meeting or rejected for specified reasons;
- users should be encouraged to take part in the discussion of each idea put forward;
- users must be confident that, if an idea is not rejected by the meeting, then action will be taken after the meeting and reported upon at the next meeting.

Unless these conditions exist, users will regard any efforts they make in maintenance and enhancement activities as a waste of time and will not play an active part.

Input to Systems Monitoring Committees

Systems monitoring committees go under a variety of names and perform a variety of functions. In general, however, they are usually composed of a mixture of systems management and staff, and user management and staff. Their purpose is usually to control systems matters pertaining to a particular area. It is possible for users to input to such committees in three ways:

- letter;
- via somebody who sits on the committee;
- by personally presenting an idea to the committee.

The conditions for successful use of departmental meetings as a change mechanism also apply here. Users will only use this mechanism effectively if they are confident that they will be taken seriously and that some action will be taken.

Direct Input to a Computerised Change System

The use of computerised change systems represents a logical step forward in the use of existing VDU-based systems.

A computerised change system would allow users to input ideas for system maintenance or enhancement immediately they have the ideas and via their own terminals. All such ideas could then be assessed by appropriate personnel. If the terminals are on-line there is clearly potential for rapid assessment of the ideas and feedback on the assessment to the user.

CONTINUING USER SUPPORT

Introduction

Generally speaking, business activities, staff and systems are constantly

changing. In these circumstances the user support network originally implemented with the system may develop shortcomings or be less effective in
some situations. Therefore throughout the life of the system, supplementary
measures should be taken, where the user support package is not meeting its
objectives.

The most common circumstances in which supplementary user support
measures may be required are:

— changes to the system;

— local experts leaving;

— new staff;

— as staff develop their system abilities and knowledge.

Changes to the System

Information about major changes to a system should be passed to users via a
formal training programme.

When minor changes are made, the following user support measures should
be taken:

— users should be informed of the change and its implications before it is
  implemented;

— users should be given the chance to ask questions about the change;

— all updated documentation should be in place when the change actually
  occurs.

A once-yearly system review may be useful in keeping users up-to-date
with their system and may also enable a clear position of future needs to
emerge. In the course of such a review all changes that have been made to the
system in the past year could be described and their cumulative impact assessed. Users could discuss their experiences — good and bad — with the system,
pass on useful tips to each other and agree what changes to the system in the
coming year would be most beneficial.

Local Experts Leaving the Organisation

The value of local experts is not often formally recognised. However their
departure can leave a large gap in the user support package. Clearly a pre-
requisite for being aware that a gap might appear is knowing who the local
experts are. If this is known, one way of filling the gap is to establish what
extra knowledge he has, ie what makes him a local expert. Another person
can then be instituted as the new local expert.

New Staff

New staff have even greater user support needs than users about to convert to a new computer system, since they not only have a new system to contend with but a new job in a new organisation with new people. However, despite this, new staff often get less support than the users of a new computer system. The minimum amount of user support that should be given to new staff is:

— training in the broader aspects of the system, its purpose and its interfaces with other systems;

— acquaintance with the terminal equipment, control equipment, computing mechanisms and processing unit which comprise the computer hardware. This should include seeing the equipment and getting a broad idea of how it works.

— training in —

  — what the system can do for him in his job;

  — how to use the system and equipment;

  — what to do when there is a systems failure.

— training in how to properly adjust his workstation and equipment to suit his own body dimensions;

— training in —

  — what documentation is available, where it is kept and how to use it;

  — what human support is available;

  — what change mechanisms are available and how to use them.

— the application of the organisation's policy on eyesight.

Where feasible, a comprehensive training programme should be devised for new starters.

The Development of Staff Capabilities

Once the initial training needs of a user have been met, it is very tempting to assume that the training function has been fulfilled. In reality, some users vastly exceed the expectations of systems staff in the level of knowledge they acquire, while others do not learn the simplest operating procedures. The differences between expected and actual levels of user expertise make it clear that users are not passive recipients of a specified, predefined amount of training. Some users actively resist acquiring the most simple skills because they see them as inconsistent with their work-role and status, while others wish to acquire far more expertise in using the system than was ever envisaged to be relevant to their work-role. In effect this means that there will be a continued training requirement, although it is unlikely to be apparent if the system is operating satisfactorily. Probably the only way of establishing what the requirement is, is to talk to the users about it. Once the requirement has been established, the most appropriate training method should be used. This may merely involve informal instruction from a more experienced user (ie the local expert) or may require a full-blown training session.

There is one aspect of a computer system in which it is prudent to give users constant refresher training; the procedures for handling faults and breakdowns. Carrying out actual trial simulations of such occurrences and undertaking constructive inquests into what happened is an effective way of ensuring that users know what to do in such situations.

Appendices

1 Checklists

1*.1 SETTING UP LINES OF COMMUNICATION

1 List the people in the following categories who will need to be kept informed of the development and its progress:

— users;

— management;

— unions;

— user associates.

2 Identify the information needs of each of the above categories of people.

3 Decide on the methods by which the information will be disseminated and feedback can be obtained.

4 Make one or more individuals responsible for information dissemination over the life of the development.

The first digit indicates the chapter to which reference should be made.

1.2 PLANNING THE DESIGN STRATEGY

1 Become acquainted with the spectrum of design strategies that are available by reading the section on design strategy.

2 Identify:

— the management who will control the development;

— the potential users of the proposed system;

— the designers of the proposed system.

3   Interview users, managers and designers to establish what experience they have had of system developments, systems design work and design strategies. Use the interviews as an opportunity to get familiar with the work environment and to establish a rapport with the users.

4   Decide on a particular strategy to obtain constructive user contributions to the design, bearing in mind the benefits and problems involved.

5   Plan any necessary training to allow users, designers and management to follow the chosen strategy.

6   Inform all parties of the strategy that is to be employed and ensure that they each understand the roles that they will have to play.

1.3 PLANNING FOR DISRUPTION

1   Identify any disruptions that are likely to occur in the following categories:
   - management control activities;
   - user contributions to systems design;
   - physical changes to the workplace;
   - installation of equipment and furniture;
   - training the users to operate the new system;
   - system testing, parallel running and system initiation.

2   Plan to cater for the disruptions identified in 1 above.

3   Ensure that all potential disruptions are understood and catered for.

1.4 WORK SITUATION APPRAISAL

1   Familiarise yourself with the user's work environment, ie gain an appreciation of the nature of the work, the pressures, priorities, skills, values, and attitudes of the people.

2   Identify the specific human factors information to be collected.

3   Review the available data collection techniques and select those likely to be appropriate and acceptable.

4   Identify the routes to follow through the organisational structure in the detailed analysis phase.

2.1 USER IDENTIFICATION

1   Identify potential users by identifying the name and job titles of the users

of the existing system.

2 Develop a 'map' of the user population. For example, it will be helpful to classify future users in terms of primary, secondary or tertiary users to avoid focussing solely upon the main users at the expense of the others.

3 Use the completed 'map' as a directory of potential users when user needs and characteristics have to be established at various points in the development.

2.2 TASK DEMANDS

1 Identify the range and functions undertaken by the users with particular emphasis on the variations in the work.

2 Review techniques for task analysis and select appropriate techniques.

3 Conduct detailed analyses of user tasks with emphasis on task differences.

2.3 WORK ROLE ANALYSIS

1 Gain an understanding of the user's perception of his work, its purpose, value and meaning for *him*.

2 Gain the user's views on the rewards and costs of the job.

3 Identify the most important aspects of the job for the user.

2.4 USER READINESS

Assess the user's receptivity/state of readiness for the proposed system. For example:

1 Assess user's previous experience of computing.

2 Examine user attitudes to computer use.

3 Assess user knowledge of the proposed computer application.

4 Assess user's willingness to cooperate in system development.

3.1 THE MECHANICS OF THE SELECTION PROCESS

1 Establish a small committee which will have responsibility for items 2 to 5 on this checklist.

2 Draw up lists of:

— functional criteria;

— general criteria;

— human aspects criteria.

3 Identify essential criteria.

4 Assign weighting to each of the identified criteria.

5 Identify the factors on which each criterion is to be judged; then assign a scoring system to the factors.

6 Establish which companies manufacture the equipment you require and contact them. Inspect their models and discuss your requirements. Eliminate those companies whose equipment does not meet the essential criteria. Draw up a shortlist of equipment for final evaluation.

7 Evaluate each model on the shortlist by the weighted ranking method and select the equipment on the basis of the results.

4.1 JOB DESIGN

1 Identify all tasks, old and new, which will undergo change as a result of the implementation of the computer system.

2 Guard against unchallenged assumptions about the manner in which tasks are combined to produce jobs. In particular guard against too much task specialisation, especially with respect to data input.

3 Generate alternative ways of allocating tasks to jobs by considering:

   − approaches such as job rotation, job enrichment and autonomous work groups;

   − allocations by function, product or customer;

   − the implications of the technical system.

4 Provide opportunities for user groups to explore the consequences of the alternatives by embodying the ideas in trial systems or simulations, and assist the process of determining a job design solution.

5 Retain flexibility in the implemented system in order that user groups can continue to adjust the allocation of tasks to jobs to suit emerging demands.

4.2 ACTIVITY FEEDBACK

1 Identify the individuals for whom activity feedback is to be provided. Establish what tasks they will be performing.

2 Decide the activity feedback information to be provided for each individual. Discuss the choice of information with the individual concerned and ensure that he feels the activities chosen are significant parts of his job.

3 Consider how the activity feedback information selected in 2 relates to existing management control information.

4 Design the details of the activity feedback information such as:

- how the information will be collected;

- how it is to be presented;

- the timing of the information.

5 Ensure that the individual for whom the information is to be provided knows:

- how the information will be collected;

- how the information is analysed;

- how to interpret the information when he receives it.

4.3 DIALOGUE DESIGN

1 Obtain the following information:

- details of the activities to be undertaken by the users;

- roles that the dialogue users play in the working situation;

- technical jargon, abbreviations, codes or cultural linguistic differences that are specific to the users;

- details of the equipment to be used to conduct the dialogue;

- details of dialogue design standards already existing in the organisation, and details of any dialogues which the users already encounter on a regular basis.

2 Remember the following general principles:

- plan to put a lot of effort into dialogue design;

- apply dialogue design standards;

- design a flexible dialogue;

- ensure that the dialogue is logical and easy to use from the user's point of view.

3 Identify the data elements that are likely to occur in the dialogue and define their characteristics.

4 Select a dialogue style.

5 Break down the dialogue into discrete blocks produced alternately by the computer system then by the user.

6 Design the dialogue for each block, for users of intermediate capability.

7 Design validation procedures and error messages.

8 Design the screen formats.

9 Test and modify the design.

10 Design flexibility into the dialogue to accommodate new users, very inexperienced users, very sophisticated users, and/or other categories of
   user.

4.4 DESIGNING PROCEDURES FOR FAULTS AND BREAKDOWNS

1 Design procedures for rectifying:

   — system faults;

   — system breakdowns;

   — terminal faults;

   — terminal breakdowns.

2 Design alternative procedures for business operations when:

   — a system breakdown occurs;

   — a terminal breakdown occurs.

3 Design procedures such that users can take positive action when they are
   confronted by a fault or breakdown. Ensure that users are trained in
   these procedures.

4 Design procedures for the recording of faults and breakdowns.

5 Ask users for their constructive criticism of all of the above procedures
   which will affect them. Where possible ask them to test the procedures.
   Make any necessary modifications.

5.1 COLLECTION OF ENVIRONMENTAL DATA

1 Get agreement from user management to carry out a preliminary survey
   of the workplace. Ensure that user management informs the occupants of
   the workplace about the survey and its purpose before it takes place.

2 Measure the workplace. Use the results to produce a small-scale plan of
   the workplace which is suitable for photocopying.

3 Return to the workplace with 6 copies of the small-scale plan. Mark
   details in rough of each of the following on separate plans:

   — layout of equipment and furniture;

   — access requirements;

   — floor characteristics;

- position of air conditioning or heating equipment;
- position and type of lighting;
- position and type of plug sockets and cable runs.

4 Check noise and vibration characteristics of all equipment.

5 After all the data has been collected, clearly record and label it on appropriate plans so that it is easy to read and understand.

5.2 WORKSTATION DESIGN

1 Investigate:

- the equipment that is to be used;
- the tasks that are to be undertaken in conjunction with the equipment;
- any other tasks that are to be undertaken at the workstation;
- the human beings who will be using the workstation;
- the layout of the workplace in which the workstation is to be located;
- the workstations currently in use.

2 Remember the three workstation design maxims (see pages 76–77). Decide on the following basic design parameters:

- the user population;
- sitting or standing workstations;
- heights;
- viewing distances;
- eye and head movement;
- document holder requirements.

3 Establish worksurface dimensions and layout by considering the following criteria:

- arm reach;
- size of equipment;
- handedness;
- task requirements;
- use of documents or manuals;
- display of notices;
- office space constraints.

4 Determine the characteristics of the chair that will be required. Inspect chairs available on the market until two or three options that fit the bill are found. Determine the characteristics of the footrest that will be required.

5 Establish the details of the workstation design by considering the following factors:

   — leg space;

   — storage space;

   — modularity;

   — ease of cleaning;

   — cabling;

   — materials;

   — weight;

   — mobility;

   — floor covering;

   — waste disposal facilities;

   — safety.

6 Check the design to ensure that the following activities have been catered for:

   — the setting and monitoring of the terminal controls;

   — use of the equipment;

   — maintenance of the equipment;

   — the installation and replacement of the equipment.

7 Produce a prototype workstation. Draw up a list of factors to be evaluated and the procedures for evaluating each factor. Evaluate the workstation under as realistic conditions as possible. Ensure that any muscular strain, visual fatigue or other health and safety hazards are identified. Modify the design and evaluate again.

   Continue this process until a satisfactory design is obtained.

5.3 LIGHTING

1 Ask the occupants if they think the lighting in the room is satisfactory.

2 Obtain the information about activities established at the job design stage.

3 Consult the IES code to establish:

   — recommended illuminance levels;

   — a suitable colour appearance for the lamps for the particular activities to be undertaken in the room.

4 Measure existing luminance levels in the room, and determine the colour appearance of the existing lamps. Match these against the requirements established in 3 above. Decide if new lamps are required. If so, ensure that the lamps that are purchased will provide the required illuminance levels when fitted with the appropriate glare shielding.

5 Establish if a glare problem exists in the room. If so, fit appropriate glare shielding. Ensure that any glare shielding fitted to the lamps does not reduce illuminance below the required levels.

6 Ensure that the reflection levels in the room are satisfactory.

7 If any changes have been made as a result of following the above steps, ask the occupants if they think the lighting in the room is now satisfactory.

8 Record the new position of lamps and/or local lighting on the scale plan of the room.

5.4 ROOM CLIMATE

1 Investigate the following physical characteristics of the room:

   — the heating system;

   — doors;

   — windows;

   — radiant temperatures.

Also record the room temperature and outside temperature (morning, midday and evening) for a period of two weeks.

2 If any equipment is to be installed in the room, inspect it in operation before installation and gauge the amount of heat output and the amount of air flow generated.

3 Spend some time in the room. Ask the occupants:

   — if the room sometimes gets too hot;

   — if the room sometimes gets too cold;

   — if the room sometimes gets too stuffy;

   — if there are draughts — if so where do they come from;

       — what they think of the heating and ventilation conditions generally.

4   Reach a decision with the appropriate management about what changes, if any, are required to the room climate. This decision will probably take the form of either doing nothing, or instituting minor changes (such as installing window blinds or getting the heating system overhauled) or of calling in a heating engineer to undertake more detailed work.

5   If any work is carried out on the room climate, visit the room about 1 month afterwards and ask the occupants what they think of the new conditions. Make three other similar visits — each visit corresponding with the middle of a season (room climates need to be checked in all of the seasons of the year). Decide if further work is required. If so, carry it out and continue the checks until a satisfactory room climate is obtained.

5.5   NOISE

1   Spend some time in the room. List the noises that can be heard and identify their sources. Ask the occupants' views on noise levels in the office.

2   If equipment is to be used in the office, listen to it in use before it is installed. Consider the noise level figures quoted by the manufacturer. Decide if it worth trying to purchase a silencer or reduce the noise output in some other way. If so, pursue that course of action.

3   Assess the general level of noise in the environment. If it is too high, consider the following possible remedies:

       — increasing the sound absorption capabilities of the room with such things as carpeting and sound absorption room dividers;

       — reducing the levels of other sources of noise in the room such as telephone bells;

       — calling in a noise engineer.

4   If noise reduction work has been undertaken, or if new equipment has been installed in the office, review the noise conditions after a month or so. Ask the occupants of the room what they think of the noise conditions in which they work. Decide if further work is required. If so, carry it out and continue to review the office noise conditions until they are satisfactory.

5.6   ROOM LAYOUT

1   Establish:

       — if any work groups are to be kept physically together;

       — if new work groups are to be created;

    — if some people do not want to be located close to others (eg, smokers and non-smokers);

    — if some people need to be located close to others to facilitate the work to be done.

2 Determine what sort of office is to be set up:

    — open plan;

    — individual offices;

    — mixture of both.

If either of the latter two are decided upon, establish which personnel are to be located in which type of office.

3 Establish what space standards, if any, are to apply. List all furniture and equipment that are to be included in each room. Calculate the space availability. If it is unsatisfactory, change the numbers of people or contents of the room until the space availability is satisfactory.

4 Use the general room information obtained during the collection of environmental data to draw a large-scale plan of the room. Ensure that doors, windows and power and communication points are all marked on the plan. Or, prepare a link diagram, using information on the connections between people and people, and people and equipment.

5 Shade in those areas of the room which are unsuitable for siting workstations. Identify such positions by considering the following constraints on the workstation positioning:

    — direction of light;

    — avoiding glare;

    — avoiding reflections on VDU screens;

    — avoiding positions subject to excessive heat or draughts;

    — avoiding positions subject to excessive noise;

    — position and size of major passageways;

    — door access requirements;

    — window access requirements;

    — equipment maintenance access requirements;

    — fire-fighting equipment access requirements.

6 Make scale cutouts to represent the furniture, workstations and equipment to be fitted into the room. Try out various arrangements of the cut-

outs on the plan. When possible alternatives are established, evaluate them against:

— work group requirements;

— constraints on the positioning of workstations;

— health and safety criteria for the routeing of cabling.

7 Present the possible layouts to the potential occupants and ask them for their comments.

8 Make a final decision about the layout that is to be implemented, bearing in mind the views of the occupants. Plan to implement the room layout taking care to cater for the disruption caused to the occupants.

Implement the layout.

9 Establish identifiers for each piece of equipment and label the equipment accordingly. Make an entry for each terminal on a Terminal Identification Record.

10 Review the layout after it has been in operation for 2 or 3 months and make any necessary changes.

5.7 MAINTENANCE CHECKS

Equipment

1 Is the screen clean? If not, clean it.

2 Is the image clear? If not, record details.

3 Is the image stable? If not, record details.

4 Are the keys clean? If not, clean them.

5 Are the top, sides and underneath the equipment clean? If not, clean them.

6 Any complaints about response times? If so, record details.

Workstation

7 Is there sufficient space for papers and reference documents on the workstation surface? If not, record details.

8 Is the workstation surface tidy? If not, tidy it.

9 Are the workstation drawers tidy? If not, tidy them.

10 Can the drawers be opened easily? If not, record details.

11 Is the general condition of the workstation desk satisfactory? If not,

record details.

12   Is the general condition of the workstation chair satisfactory? If not, record details.

13   Does the chair adjusting mechanism work satisfactorily? If not, record details.

14   Do all the users know how to adjust the chair? If not, record that training is required.

15   Are the wastepaper facilities adequate? If not, record details.

16   Is the condition of the notices displayed on the workstation satisfactory? If not, record details.

17   Is it easy to get in and out of the workstation? If not, record details.

18   Is the condition of the footstool satisfactory? If not, record details.

19   Is any wiring cluttering up the workstation? If so, record details.

Lighting

20   Are any of the room lights not working? If so, record details.

21   Are any of the fluorescent room lights flickering? If so, record details.

22   Are there any sources of glare in the room? If so, record details.

23   Do any of the lights, lampshades or diffusers need cleaning? If so, record details.

24   Are the window blinds easy to adjust? If not, record details.

25   Do the window blinds effectively shield the occupants from the sun's rays and from glare? If not, record details.

Room Climate

26   Is the room sometimes too hot or too cold? If so, record details.

27   Are there any draughts in the room? If so, record details.

28   Is the room affected by dust, fumes or smells? If so, record details.

29   Does the room get stuffy? If so, record details.

Noise

30   Is the room too noisy? If so, record details.

Room Layout

31   Is there adequate unobstructed passageway space in the room? If not,

record details.

32   Does any wiring cross any passageways? If so, remove it.

33   Does any wiring obstruct the feet in the course of using the workstation? If so, record details.

34   Is any of the wiring severed or frayed? If so, take any dangerous wiring out of service and record details.

35   Are there any broken or loose plugs or sockets? If so, take any dangerous plugs or sockets out of service and record details.

5.8 UPKEEP

1   Allocate responsibility for the upkeep of equipment, workstations and the workplace to a specific job title.

2   Identify the checks that must be made, or work that must be carried out, to undertake the planned maintenance activity.

3   Specify the time period between planned maintenance inspections for each check.

4   Establish who is to carry out each maintenance check.

5   Establish the procedure for obtaining the view of the users when planned maintenance is carried out.

6   Establish procedures for undertaking each item of planned maintenance and for reporting on planned or breakdown maintenance activity.

7   Draw up a planned maintenance schedule and record it on a suitable planning chart.

8   Establish procedures for monitoring maintenance records to identify any inherent weaknesses or likely future breakdowns.

6.1 PLANNING USER SUPPORT

1   Appoint one person to be responsible for the design and implementation of the user support network.

2   Establish:

  — the users of the new system and the relevant skills or knowledge they possess;

  — what new tasks the users will have to perform;

  — details of existing human support mechanisms;

  — details of existing documentary support;

    — details of existing change mechanisms.

3 Plan human and documentary support.

4 Plan change mechanisms.

5 Plan the training programme.

6 Review the support network and make modifications where necessary.

7 Implement the support network.

6.2 TRAINING

1 Identify all the people who will be affected by the new system.

2 Establish the training needs of each individual.

3 Determine teaching objectives for each need established.

4 Identify appropriate training modules.

5 Determine what techniques are to be used to carry out each module. Plan the content of each module.

6 Arrange times, dates, places and people for the programme to be actually carried out. Ensure that:

    — the trainees understand why they are undertaking the training programme;

    — user management has set aside appropriate time for staff to undertake the programme.

7 Carry out the programme, monitor and control it to ensure that the objectives are achieved within time and resource constraints.

8 Review the training programme. Identify:

    — any improvements that could have been made;

    — any outstanding or future training requirements.

Feed this information into the next training programme.

6.3 DOCUMENTARY SUPPORT

1 If terminal equipment is being installed obtain the following documents from the terminal manufacturers:

    — installation instructions;

    — maintenance instructions;

    — the technical specification of the terminal;

— the terminal operation guide.

Pass the installation and maintenance instructions, and the technical specification to the job title responsible for maintenance activities.

2 Establish what documentation currently exists in the organisation. Consider the job of each user in turn. Decide whether he will require a user's guide to equipment operation, a user's guide to the system or any other documentation.

3 Decide which of the following methods of providing documentary support is to be used for the user documentation:

— paper-based manual;

— paper-based circulars;

— expert system;

— help facilities;

— computer-based documentation.

4 Use your own internal standards for producing the documentation. If no standards exist, decide on the format, style, presentation and method of update to be used. Get the user to assist in producing the documentation or, failing this, to vet it as it is produced.

5 Get some of the potential users to try out the documentation and note their reactions and comments. Make any necessary modifications and try it out on the users again. Continue this process until the documentation is satisfactory.

6 Set up procedures that will enable users to suggest changes to the documentation. Display these procedures in a prominent position, either in the documentation or on the documentation workstation.

7 Establish which job title has responsibility for making amendments to the documentation.

8 For paper-based documentation, establish which job title has responsibility for the physical update of the manuals.

6.4 HUMAN SUPPORT

1 Read about the various types of human support mechanisms in the relevant section of this book.

2 Establish what human support mechanisms already exist within the organisation. Determine how successful they are. Consider the activities each user will be carrying out. Decide what human support, if any, he should be provided with.

3 Set up the human support mechanisms required. Ensure that the support person:

   — fully understands the job he has to do;

   — is capable of carrying out the job;

   — is formally recognised to be carrying out the support function.

4 Ensure that the users concerned know what human support is available for their use.

5 Review the human support mechanisms after their first 6 months of use. Seek user opinion. Ask the people providing the support how well they think the mechanism is working. Make any appropriate modifications.

6.5 CHANGE MECHANISMS

1 Read about the various types of change mechanisms in the text of this section.

2 Determine what change mechanisms currently exist and how successful they are.

3 Decide what change mechanisms will be most appropriate for each user.

4 Set up the change mechanisms and ensure that appropriate feedback mechanisms are also instituted.

5 Review the change mechanisms after their first six months of use. Seek user opinion. Make any appropriate modifications.

6.6 CONTINUING USER SUPPORT

1 Ensure that an appropriate amount of training is given to users when changes are made to the system.

2 Establish who the local experts are. Find out just what knowledge and abilities they have that make them local experts. If they are about to leave the organisation find a means of making that knowledge and capability easily available to users until a new local expert emerges.

3 Draw up a training programme for new starters. Ensure that the programme includes the following:

   — training in the broad aspects of the system and its interfaces;

   — hardware acquaintance training;

   — skills and detailed system training;

   — training in how to adjust workstation chair and workstation equipment;

— training in what documentary and human support is available;

— training in what change mechanisms are available.

Apply the organisation's policy on eyesight to the new starter. Be aware that there is likely to be a continuing training need. Meet the needs by the most appropriate training mechanisms.

2 Human Aspects Criteria (VDUs and Keyboards)

SCREENS

Character Formation

a Character height should be greater than or equal to 3 mm

b The resolution of the dot matrix must be a minimum of 5 x 7 and preferably 7 x 9

c User- and lower-case character width should be 70—80% of the upper-case character height

d Stroke width should be between 12% and 17% of the upper-case character height

e The space between the characters should be between 20% and 50% of the upper-case character height

f Row spacing should be between 100% and 150% of the upper-case character height

g For lower-case characters, descenders should project below the base line of the matrix

h It should be possible to distinguish between the following characters:

Mutual	*One-way*
O and Q	C called G
T and Y	D called B
S and 5	H called M or N
I and L	J, T called I
X and K	K called R
I and I*	2 called Z*
	B called R, S, or 8*

*These three often comprise 50% or more of the total confusions.

i It should be possible to distinguish between the number 0 and the letter O. (It should be noted that the letter 'Ø' is included in several Nordic alphabets and should not be used to represent the number 0)

j The characters should be upright, not slanted

k If filters are used, the character image should still appear sharp and well-defined

l The displayed character image should be stable

Coding Format

a It should be possible to distinguish between different luminance levels (used for selective brightening) at maximum setting

b A cursor should be provided

c The cursor should be clearly distinguishable from the other symbols on the display

d If it is possible to blink selected parts of the screen the blink rate should be between 2 and 4Hz

e If the cursor has a blink action it should be suppressible

f All displayed symbols should be unambiguous

The Display Screen and Luminance

a The character luminance should be a minimum of 45 cd/m² or preferably between 80 and 160 cd/m²

b Character luminance should be easily adjustable

c Character image should remain sharply defined at maximum character luminance

d Background screen luminance should be between 15 and 20 cd/m² under normal office lighting conditions

e The background screen luminance should be adjustable

f The contrast between character and background should be a minimum of 3:1 and preferably 8:1

KEYBOARDS

General Criteria

a The keyboard should be detachable from the screen, ie joined by a cable

b The keyboard should be heavy enough to ensure stability against unintentional movement

c   The thickness of the keyboard (ie base to the home row of keys) should be a maximum of 50 mm and preferably 30 mm

d   The angle of the keyboard should be in the range 5–15°

e   The surface of the keyboard surround should be matt finished

f   The reflectance of the whole keyboard surface (not single keys) should be between 0.40 and 0.60

g   When keystrokes are no longer being registered, eg when memory is full, a warning should be provided

Key Characteristics

a   Key pressure should be between 0.25 and 1.5 Newtons

b   Key travel should be between 0.8 and 4.8 mm

c   For square keytops the length of the diagonal should be between 12 and 15 mm

d   Centre spacing between adjacent keys should be between 18 and 20 mm

e   Key legends should be resistant to wear and abrasion, ie the legends should be moulded into the keytop

f   The keytop surfaces should be concave

g   Reflections from the keytop surfaces should be kept to a minimum

h   The activation of each key should be accompanied by a feedback signal such as an audible click, a tactile click or a snap action

j   If two keys are activated simultaneously a warning signal should be given

k   The keyboard should be provided with an n-key roll-over facility

Keyboard Layout

a   The layout of the alpha keys should correspond to that of a conventional typewriter keyboard layout

b   The layout of the numeric keys above the alpha keys should correspond to the conventional typewriter keyboard layout

c   If an auxiliary numeric keyset is provided, the keys should be arranged in the same way as either the calculator layout (ie 7, 8, 9 along the top row) or preferably the telephone layout (ie 1, 2, 3 along the top row)

d   The space bar should be at the bottom of the keyboard

e The colour of the alphanumeric keys should be neutral (eg beige or grey), as opposed to black, white or one of the spectral colours such as red, yellow, green or blue

f The different function key blocks should be distinct from the other keys by either colour, shape, or position

g All keys for which unintentional or accidental operation may have serious consequences should be especially secure by either their position, a higher required key pressure, a key lock or by a two-handed or two-key chord operation

h The function key labels and symbols should correspond to the same functions on other keyboards in the same workplace

3 Completed Data Definition Sheet (example)

Data Definition Sheet		Fill this sheet in according to the instructions below
1 Name given to data element by user	back order	Enter required information
2 User abbreviation of 1)	B/O	Enter
3 Definition of 1)	An order against stock which cannot be met until a replenishment quantity is received.	Enter
4 Coded/not coded/mixture of coded and uncoded		Underline the one that is applicable
5 Arithmetic/Not arithmetic		Underline the one that is applicable
6 Alpha/numeric/alphanumeric/other		Underline the one that is applicable
7 Length	7	Enter maximum (count decimal point as 1 space)
8 Structure	whole numbers only.	Enter
9 Value range	0 - 9999999	Enter if applicable: If not enter N/A
10 Units of measurement	Items	Enter if applicable: If not enter N/A
11 Validity checks	Numeric: in range 0 - 9999999: must be associated with a Finished Product No.	Enter
12 Responsibility for generation	Warehouse Supervisor	Enter
13 Responsibility for maintenance	warehouse Supervisor	Enter
14 Responsibility for authorising use of the data	N/A	Enter if applicable: If not enter N/A
15 Functions authorised to use the data	All	Enter
16 Timing of re-use	N/A	Enter if applicable: If not enter N/A
17 Systems in which used	Finished stock system.	Enter
18 Relationship with other data	must be associated with a Finished Product No.	Either enter details or 'none'

4 VDU/Keyboard Assessment for Workstation Design (example)

Equipment	Visual Display Unit (VDU)
CONTROLS	
1 On/off switch	A switch located at the left bottom front of the VDU
2 On/off monitor	A red light signifies on. Located next to the on/off switch
3 Key depression feedback	A dial, located at the extreme left of the keyboard, controls sound feedback
4 Brightness control	A dial located at the bottom back of the VDU

1, 2 and 3 can be used from the normal positions adopted when using the VDU. However, item 4 — the Brightness Control — requires access from either the side or the back of the unit. Because the only feedback about adjustments made to the brightness control is the brightness of the screen, it may be necessary to move from the side or back of the screen to the front and back again several times to obtain a satisfactory result. Space around the workstation must be available to do this.

OPERATION Two activities must be carried out in conjunction with one another in order to operate the VDU: Keying and Viewing.

1 The KEYING activity is to be carried out on a standard QWERTY keyboard. For this operation the arms should be unrestrained and the forearms should be at right angles to the upper arms. This position allows free movement across the keyboard and depression of the keys from the forearm as well as from the wrists and fingers.

2 Maximum and minimum VIEWING distances for displays are generally accepted to be in the range of 71 and 33 cm respectively with an optimum in the region of 45–50 cm. The eye is susceptible to strain if it is asked to continually refocus over a period of time. Therefore the distances between the eye and the screen and the eye and the keyboard should not only be

within the limits described above, but should also be kept as similar as possible.

MAINTENANCE Access to the back of the unit is required for maintenance work. Therefore the unit must be either freestanding so that it can be
turned round on the surface it is standing on, or else there must be adequate
space behind the terminal in which the maintenance engineer can stand and
take things in and out of the unit cabinet.

INSTALLATION/REPLACEMENT The unit can be carried by one person
and has no particular features which would cause difficulties in the installation
or replacement operation.

SERVICES The unit requires a standard mains power source and a communications cable. Both cables are attached to the back of the unit and
should be led away from the workstation via a route so that the user will not
come into contact with them. The keyboard is linked to the screen unit via a
1 metre long cable.

SIZE/WEIGHT/STABILITY The VDU screen unit measures 46 cm wide,
40 cm long and 32 cm high. It weighs 31 kilos and rests in a stable condition
on four small anti-slip stops, one at each corner. The keyboard is 46 cm long,
27 cm wide and 8 cm high (at its highest point). It weighs 7 kilos and rests in
a stable condition on four small anti-slip stops, one at each corner.

5 Manuscript Holder Recommendations for VDU Tasks

Task	Frequency of manuscript change	Type and position of manuscript holder for right-hand holders
Pure copy entry, no manipulation	Seldom	Size: according to type of document. Row marker: yes. Position: to the left of the display screen, 20 degree angle.
Copy entry with some manipulation, eg occasional notes	Seldom	Size: according to type of document. Row marker: yes. Position: to the right of the display screen.
Pure copy entry, no manipulation	Frequent	Size: according to type of document. Row marker: preferably. Position: to the left of the display screen, or between the keyboard and the screen.
Copy entry with some manipulation	Frequent	Size: according to type of document. Row marker: preferably. Position: to the right of the display screen.
Pure copy entry, no manipulation (mostly numerical data)	Very frequent	Size: according to type of document. Row marker: no. Position: to the left of the keyboard.

6 VDU Workstation Design Factors

Factor to be evaluated . . . . .Optimum state of factor

1 Worksurface height . . . . .adjustable 52–67 cm

2 Seat height. . . . . .adjustable 34–52 cm

3 Viewing distances . . . . .45–50 cm

4 Neck/head movement. . . . . .Should not be excessive

5 Document holder facilities . . . . .Should be easy to use

6 Arm reach . . . . .All items of equipment and job aids which must be frequently manually manipulated by the user, should be within the normal arm reach of the user, ie within reach without requiring movement of the body

7 Positioning of job aids and . . . . .Should be sited according to:
   items of equipment
                                                      – their frequency of use
                                                      – their relation to the way the activity is performed
                                                      – their importance

8 Handedness . . . . .Should cater for both left- and right-handed people

9 Space for documents and . . . . . .There should be sufficient space to use documents and manuals as demanded by the job
   manuals

10 Display of notices . . . . .Notices should be eye-catching and easy to read from a normal position. They should not be obscured as a result of other activities at the workstation

11 Chair stability. . . . . .Chairs should be safe from tipping over and should preferably have a five-point base

12 Seat height adjusting . . . . .Chairs should be easy to adjust from mechanism   the seated position, and safe against self or unintentional release

13 Seat front edge . . . . .Should be rounded to avoid cutting into the thighs

14 Seat surface . . . . .Should be padded

15 Backrest . . . . .Should be adjustable forwards/backwards, up/down

16 Footrest adjustability. . . . . . . .Adjustable footrest should be provided. Should be easily and quickly adjustable in height (0–5 cm) and inclination (10°–15°)

17 Footrest surface . . . . .Should allow comfortable movement of the feet without slipping

18 Thigh clearance. . . . . .Should be a minimum of 18 cm

19 Leg area. . . . . .Should be at least 80 cm wide, 70 cm deep, and free from obstructions

20 Ease of cleaning . . . . .Should not be difficult to clean those parts that will require cleaning

21 Storage space . . . . .Should be adequate space for storage of documents, handbooks, personal belongings, etc, as required

22 Cabling . . . . .Should not:
      — interfere with workstation tasks
      — be loose on the workstation
      — look untidy
      — interfere with the legs or feet
      — be in a position where it might get severed

23 Worksurface material . . . . .Should be matt finished & have a reflectance of 0.4 preferably, or 0.6 as an absolute maximum

24 Mobility . . . . .If casters are fitted they should move easily

25 Stability . . . . .The workstation must be stable when

		in position
26	Waste disposal facilities.	Should be easy to use, easy to clear, and of adequate capacity
27	Corners/sharp edges	There should be no pointed corners or sharp edges
28	Sitting monitoring of . . . . equipment controls	Should be possible without taking up unsafe postures
29	Installation and replacement . . . . of the equipment	Should not entail a major dismantling of the workstation

7   VDU Eye Test Record Sheets

TO BE COMPLETED BY SUBJECT

Name _____ Sex _____ Date of Birth _____

Organisation _____

Job _____

Have you had a VDU Eye test before? ____If yes, when was it? _____

How long have you used a VDU? _____

Make and model of VDU used: _____

On average, how many hours per day do you use a VDU?: _____hours per day

Is this all at one time?: _____If not, how long is a typical session? _____

Tick the category that best describes your **own** use of a VDU.

☐   Input only

☐   Mainly input but reading some output also

☐   A mixture of inputting and reading output

☐   Mainly reading output but some input also

☐   Reading output only

Do you work mainly with

☐   Text, ie mainly words and alpha characters

☐   Data, ie mainly numbers

☐   Program instructions

TO BE COMPLETED BY TESTER

TEST	RIGHT EYE			LEFT EYE		
Unaided Visual Acuity						
Refractive Findings	SPH	CYL	AXIS	SPH	CYL	AXIS
Corrected Visual Acuity						
Amplitude of Accommodation (Dioptres)	Corrected for near=			Corrected for near=		
Suppression	YES/NO			YES/NO		
Muscle Balance - Distance (6m)	Horizontal					
(Maddox Rod)	Vertical					
Muscle Balance - 1 Metre	Horizontal					
(Maddox Rod)	Vertical					
Muscle Balance Near (33cm)	Horizontal					
(Maddox Wing)	Vertical					
Near Point of convergence PPC						

Spectacles
currently used   Distance _____Close   _____Work only _____Remarks:

New spectacles
required   _____   _____   _____

Optician's Signature: Address:

8 Office Space Availability Measurement (example)

MEASUREMENT OF OFFICE SPACE

USERS' GUIDE

1 Use separate calculation sheets for each enclosed office area

2 Ask permission of the occupants of the office before you start to measure

3 Draw a sketch of the area and its furniture on the calculation sheet

4 Measure up the area and furniture concerned, to the nearest centimetre, entering the lengths, widths and heights on the calculation sheets as you go (an example of a completed calculation sheet is shown overleaf)

5 When you finish taking measurements use the calculator to do the necessary calculations to an accuracy of one decimal place

DATE: 17 JAN 80

OFFICE SPACE CALCULATION SHEET

Name of Office   NEIL WALKER/JACK HARDIE

No. of people working in Office   2   **(A)**

Draw a sketch of the office in the space below. Write in the measurements
of the office perimeter on the sketch. Draw in the furniture you will be
taking account of in the survey. Number each sketch of a piece of furniture.

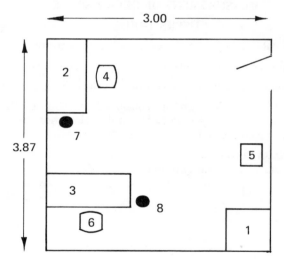

Ceiling Height   =   2.31 m

Floor Area of Office   =   11.61 sq/metres **(B)**

Volume of Office   =   26.82 sq/metres **(C)**

The numbers in the extreme left column overpage refer to the numbers
you have assigned to each of your pieces of furniture. Remember that the
area measurement should be obtained for the largest area taken up by the
piece of furniture; and that the volume measure should be obtained for the
largest effective volume taken up by the piece of furniture, ie for a chair the
height would be to the top of the chair back.

No.	DESCRIPTION OF PIECE OF FURNITURE	Measurements to obtain AREA		Area taken up	Measurements to obtain VOLUME		VOLUME TAKEN UP
		LENGTH	WIDTH		AREA	HEIGHT	
1	Filing cabinet	0.47	0.61	0.29	0.29	1.32	0.38
2	Desk	0.76	1.53	1.16	1.16	0.71	0.82
3	Desk	0.76	1.53	1.16	1.16	0.71	0.82
4	Chair for 2	0.61	0.55	0.33	0.33	0.81	0.27
5	Chair for visitors	0.49	0.49	0.24	0.24	0.76	0.18
6	Chair for 3	0.61	0.55	0.33	0.33	0.81	0.27
7	Bin	0.15		0.07	0.07	0.27	0.02
8	Bin	0.15		0.07	0.07	0.27	0.02
		TOTAL AREA =		3.65 (D)	TOTAL VOLUME =		2.78 (E)

Floor Area of Office/Person $= \dfrac{(B)}{(A)} = \dfrac{11.61}{2} = 5.81$ sq metres/person

Effective Working Area/Person $= \dfrac{(B) - (D)}{(A)} = \dfrac{11.61 - 3.65}{2} = 3.98$ sq metres/person

Volume of Office/Person $= \dfrac{(C)}{(A)} = \dfrac{26.82}{2} = 13.41$ cu metres/person

Effective Working Volume/Person $= \dfrac{(C) - (E)}{(A)} = \dfrac{26.82 - 2.78}{2} = 12.02$ cu metres/person

9 Terminal Identification Record Sheet (example)

Terminal Identification Record				
Terminal Identifier	Type of Terminal	Geographical Location	Responsibility for Documentation Maintenance	Responsibility for Physical Maintenance

10 Stock Control Clerk Training Needs (example)

- To understand in general terms the objectives of the system and how those objectives will be achieved
- To know what other systems interface with this system and how they do so
- To become acquainted with the computing operation in the organisation
- To understand that the organisation wants him to look out for opportunities to improve the system
- To know how to go about suggesting changes to the system
- To understand what changes there will be to his job and how he will benefit from the changes
- To know what records are held for each item of stock and to understand how the value or content of each record is obtained
- To know what input he may have to make to the system and how to go about it
- To know what output he may receive from the system and how to interpret it
- To know what each key on the keyboard does
- To be able to make control adjustments to the equipment (eg brightness etc)
- To be able to use the keyboard confidently
- To be able to adjust the height of the seating, position of keyboard and position of the screen to suit his own requirements
- To know what human and documentary support are available and how to use them
- To know what to do when system faults or breakdowns occur.

11 Bibliography

CHAPTER 1

Computerisation Guidelines, H F Farrow, 1979, NCC Publications, The National Computing Centre Ltd, Oxford Road, Manchester M1 7ED, England. Tel: 061-228 6333

Content: The book concentrates on the effect of a computer development on the social environment of an organisation. It is divided into five sections — each section providing advice for one of five groups of people: Management, Foremen and Supervisors, Employees and Trade Unions, Computer Specialist staff and Project Directorate.

CHAPTER 3

Computing Journal Abstracts, Published by NCC Information Services, The National Computing Centre Limited, Oxford Road, Manchester, M1 7ED, England. Tel: 061-228 6333

Content: A fortnightly bulletin of 100 abstracts of articles published in computing journals. Aspects covered include equipment, the computing industry, training, etc. A list of journals abstracted is issued annually.

The Computer Users' Year Book, CUYB Publications Ltd, 430–432 Holdenhurst Road, Bournemouth, BH8 9AA, England. Tel: 0202-302464

Content: A very comprehensive reference source to all aspects of the computer business. Updated and published annually.

Computer Review, GML Corporation, 594 Marrett Road, Lexington, Massachusetts, USA. Tel: 617-871-0515

Content: A very comprehensive reference source to computer equipment
manufacturers and the equipment they produce. Published on a
yearly basis with three mid-year updates.

Computer Hardware Record, Published by NCC Information Services, The
National Computing Centre Limited, Oxford Road, Manchester M1 7ED,
England. Tel: 061-228 6333.

Content: A monthly publication covering one category of equipment
each month.

Selection of Data Communications Equipment, Nichols & Jocelyn, 1979, NCC
Publications, The National Computing Centre Limited, Oxford Road,
Manchester M1 7ED, England. Tel: 061-228 6333.

Content: Describes the range of data communications equipment available in 1979 and details criteria by which such equipment can
be evaluated. A selection procedure is also provided.

Visual Display Terminals, A Cakir, D J Hart, T F M Stewart, John Wiley &
Sons Ltd, Barrins Lane, Chichester, Sussex, England. Tel: 0243-784531.

Content: Very comprehensive, in-depth analysis of all aspects of VDUs
and their use.

CHAPTER 4

Design of Man-Computer Dialogues, James Martin, Prentice-Hall Inc, Engle-
wood Cliffs, New Jersey, USA.

Content: A very wide ranging book covering dialogues using sound and
graphics as well as alphanumeric dialogues. Many examples are
given throughout.

Humanized Input, Tom Gilb, Gerald M Weinberg, Winthrop Publishers Inc,
17 Dunster Street, Cambridge, Massachusetts 02138.

Content: Discusses in great detail the issues to be taken into account
when designing each word or few words of an alphanumeric dialogue. Very comprehensive and quite readable.

Man/Computer Communication, Infotech State of the Art Report, edited by
Prof. B Shackel, Infotech International Ltd, Maidenhead, Berkshire,
England.

Content: This report was published in 1979 in two volumes. Volume 2 reproduces 22 papers which cover a whole range of the human aspects of computing. Volume 1 is a commentary on the state of the art in 1979 by the editor, Prof. Brian Shackel, using extracts from the 22 papers. A number of the papers deal with the subject of dialogue design.

CHAPTER 5

Interior Lighting Design, D W Durrant (editor), Lighting Industry Federation Limited and the Electricity Research Council (1973).

Content: A useful book on issues to be taken into consideration when designing the illumination of an interior.

The IES Code for Interior Lighting, Illuminating Engineering Society (London 1977) and The Chartered Institute of Building Services (CIBS), Delta House, 222 Balham High Road, Balham, London SW12 9BS.

Content: A good overview of lighting design considerations and general lighting issues.

The Ergonomics of Lighting, R G Hopkinson & J B Collins, Macdonald Technical and Scientific, London (1970).

Content: A book containing fundamental lighting knowledge; experimental studies of visual performance, glare, flicker, visual fatigue, colour of sources and adaptation, together with some applications.

Illuminating Engineering Society Technical Report No 10, (1967), Evaluation of discomfort glare.

Content: The IES Glare Index System for artificial lighting installations.

Vision and VDUs, J W Grundy and S G Rosenthal, Association of Optical Practitioners, Bridge House, 233–234 Blackfriars Road, London SE1 8NW Tel: 01-261 9661.

Content: This publication consists of:

  – 3 articles
  – a recommended standard for eye examinations
  – suggested standards of vision for VDU operators
  – a booklet for VDU operators called 'VDUs and you'

The articles are titled:

Visual Display Units — Nightmare to the Operator?;
Visual Display Units in Question; and
VDUs — Their Effects on Eyes.

Visual Aspects and Ergonomics of Visual Display Units, Documentation for a course run by the Institute of Ophthalmology, edited by Veronica M Reading, Institute of Ophthalmology, University of London, Judd St, London WC1H 9QS. Tel: 01-387 9621.

Content: This publication records the content of 13 lectures. The question of eyesight and the use of VDUs is dealt with in some scientific detail in 6 of the papers. Other subjects covered include postural stress, workstation design, glare and radiation.

Air Ions and Human Performance, by L H Hawkins and T Barker, Dept of Human Biology & Health, University of Surrey, Guildford, *Ergonomics*, Taylor and Francis Ltd, 10—14 Macklin St, London WC2B 5NF.

Noise and Man, W Burns, John Murray Ltd, 50 Albermarle St, London (1973).

Content: Noise, its measurement and its effect on hearing.

Acoustic Noise Measurements, A Bruel & Kjaer handbook, obtainable from B & K Laboratories Ltd, Bradshaw Trading Estate, Greengate, Middleton, Manchester.

Content: A technical book about the practical aspects of noise and its measurement.

Humanscale 1/2/3, Niels Diffrient, Alvin R Tilley, Joan C Bardagjy, MIT Press, Massachusetts Institute of Technology, Cambridge, Massachusetts 02142, USA.

Content: A very comprehensive selection of anthropometric measures are presented in three selector charts and an accompanying booklet. The measures are split into body measurements, link measurements, seating guide, seat/table guide, wheelchair users, handicapped and elderly.

Visual Display Terminal, A Cakir, D J Hart, T F M Stewart, John Wiley & Sons Ltd, Barrins Lane, Chichester, Sussex. Tel: 0243-784531.

Content: Very comprehensive, in-depth analysis of all aspects of VDUs and their use.

Applied Ergonomics Handbook. Reprints from Applied Ergonomics Vol 1,
Nos 1–5 and Vol 2, Nos 1–3, edited by Professor Brian Shackel, IPC
Science & Technology Press Ltd, IPC House, 32 High Street, Guildford,
Surrey, England GU1 3EW.

Content: Provides a quick simple guide to most aspects of applied ergonomics.

Index